Reviews of Environmental Contamination and Toxicology

VOLUME 209

For further volumes:
http://www.springer.com/series/398

Reviews of Environmental Contamination and Toxicology

Editor
David M. Whitacre

Editorial Board
María Fernanda Cavieres, Playa Ancha, Valparaíso, Chile • Charles P. Gerba, Tucson, Arizona, USA
John Giesy, Saskatoon, Saskatchewan, Canada • O. Hutzinger, Bayreuth, Germany
James B. Knaak, Getzville, New York, USA
James T. Stevens, Winston-Salem, North Carolina, USA
Ronald S. Tjeerdema, Davis, California, USA • Pim de Voogt, Amsterdam, The Netherlands
George W. Ware, Tucson, Arizona, USA

Founding Editor
Francis A. Gunther

VOLUME 209

Coordinating Board of Editors

DR. DAVID M. WHITACRE, *Editor*
Reviews of Environmental Contamination and Toxicology

5115 Bunch Road
Summerfield, North Carolina 27358, USA
(336) 634-2131 (PHONE and FAX)
E-mail: dwhitacre@triad.rr.com

DR. HERBERT N. NIGG, *Editor*
Bulletin of Environmental Contamination and Toxicology

University of Florida
700 Experiment Station Road
Lake Alfred, Florida 33850, USA
(863) 956-1151; FAX (941) 956-4631
E-mail: hnn@LAL.UFL.edu

DR. DANIEL R. DOERGE, *Editor*
Archives of Environmental Contamination and Toxicology

7719 12th Street
Paron, Arkansas 72122, USA
(501) 821-1147; FAX (501) 821-1146
E-mail: AECT_editor@earthlink.net

ISSN 0179-5953
ISBN 978-1-4419-6882-1   e-ISBN 978-1-4419-6883-8
DOI 10.1007/978-1-4419-6883-8
Springer New York Dordrecht Heidelberg London

© Springer Science+Business Media, LLC 2010
All rights reserved. This work may not be translated or copied in whole or in part without the written permission of the publisher (Springer Science+Business Media, LLC, 233 Spring Street, New York, NY 10013, USA), except for brief excerpts in connection with reviews or scholarly analysis. Use in connection with any form of information storage and retrieval, electronic adaptation, computer software, or by similar or dissimilar methodology now known or hereafter developed is forbidden.
The use in this publication of trade names, trademarks, service marks, and similar terms, even if they are not identified as such, is not to be taken as an expression of opinion as to whether or not they are subject to proprietary rights.

Printed on acid-free paper

Springer is part of Springer Science+Business Media (www.springer.com)

Foreword

International concern in scientific, industrial, and governmental communities over traces of xenobiotics in foods and in both abiotic and biotic environments has justified the present triumvirate of specialized publications in this field: comprehensive reviews, rapidly published research papers and progress reports, and archival documentations. These three international publications are integrated and scheduled to provide the coherency essential for nonduplicative and current progress in a field as dynamic and complex as environmental contamination and toxicology. This series is reserved exclusively for the diversified literature on "toxic" chemicals in our food, our feeds, our homes, our recreational and working surroundings, our domestic animals, our wildlife, and ourselves. Tremendous efforts worldwide have been mobilized to evaluate the nature, presence, magnitude, fate, and toxicology of the chemicals loosed upon the Earth. Among the sequelae of this broad new emphasis is an undeniable need for an articulated set of authoritative publications, where one can find the latest important world literature produced by these emerging areas of science together with documentation of pertinent ancillary legislation.

Research directors and legislative or administrative advisers do not have the time to scan the escalating number of technical publications that may contain articles important to current responsibility. Rather, these individuals need the background provided by detailed reviews and the assurance that the latest information is made available to them, all with minimal literature searching. Similarly, the scientist assigned or attracted to a new problem is required to glean all literature pertinent to the task, to publish new developments or important new experimental details quickly, to inform others of findings that might alter their own efforts, and eventually to publish all his/her supporting data and conclusions for archival purposes.

In the fields of environmental contamination and toxicology, the sum of these concerns and responsibilities is decisively addressed by the uniform, encompassing, and timely publication format of the Springer triumvirate:

> *Reviews of Environmental Contamination and Toxicology* [Vol. 1 through 97 (1962–1986) as Residue Reviews] for detailed review articles concerned with any aspects of chemical contaminants, including pesticides, in the total environment with toxicological considerations and consequences.
>
> *Bulletin of Environmental Contamination and Toxicology* (Vol. 1 in 1966) for rapid publication of short reports of significant advances and discoveries in the fields of air, soil,

water, and food contamination and pollution as well as methodology and other disciplines concerned with the introduction, presence, and effects of toxicants in the total environment.

Archives of Environmental Contamination and Toxicology (Vol. 1 in 1973) for important complete articles emphasizing and describing original experimental or theoretical research work pertaining to the scientific aspects of chemical contaminants in the environment.

Manuscripts for Reviews and the Archives are in identical formats and are peer reviewed by scientists in the field for adequacy and value; manuscripts for the *Bulletin* are also reviewed, but are published by photo-offset from camera-ready copy to provide the latest results with minimum delay. The individual editors of these three publications comprise the joint Coordinating Board of Editors with referral within the board of manuscripts submitted to one publication but deemed by major emphasis or length more suitable for one of the others.

Coordinating Board of Editors

Preface

The role of *Reviews* is to publish detailed scientific review articles on all aspects of environmental contamination and associated toxicological consequences. Such articles facilitate the often complex task of accessing and interpreting cogent scientific data within the confines of one or more closely related research fields.

In the nearly 50 years since *Reviews of Environmental Contamination and Toxicology* (formerly *Residue Reviews*) was first published, the number, scope, and complexity of environmental pollution incidents have grown unabated. During this entire period, the emphasis has been on publishing articles that address the presence and toxicity of environmental contaminants. New research is published each year on a myriad of environmental pollution issues facing people worldwide. This fact, and the routine discovery and reporting of new environmental contamination cases, creates an increasingly important function for *Reviews*.

The staggering volume of scientific literature demands remedy by which data can be synthesized and made available to readers in an abridged form. *Reviews* addresses this need and provides detailed reviews worldwide to key scientists and science or policy administrators, whether employed by government, universities, or the private sector.

There is a panoply of environmental issues and concerns on which many scientists have focused their research in past years. The scope of this list is quite broad, encompassing environmental events globally that affect marine and terrestrial ecosystems; biotic and abiotic environments; impacts on plants, humans, and wildlife; and pollutants, both chemical and radioactive; as well as the ravages of environmental disease in virtually all environmental media (soil, water, air). New or enhanced safety and environmental concerns have emerged in the last decade to be added to incidents covered by the media, studied by scientists, and addressed by governmental and private institutions. Among these are events so striking that they are creating a paradigm shift. Two in particular are at the center of ever-increasing media as well as scientific attention: bioterrorism and global warming. Unfortunately, these very worrisome issues are now superimposed on the already extensive list of ongoing environmental challenges.

The ultimate role of publishing scientific research is to enhance understanding of the environment in ways that allow the public to be better informed. The term "informed public" as used by Thomas Jefferson in the age of enlightenment conveyed the thought of soundness and good judgment. In the modern sense, being "well informed" has the narrower meaning of having access to sufficient information. Because the public still gets most of its information on science and technology from TV news and reports, the role for scientists as interpreters and brokers of scientific information to the public will grow rather than diminish. Environmentalism is the newest global political force, resulting in the emergence of multinational consortia to control pollution and the evolution of the environmental ethic. Will the new politics of the twenty-first century involve a consortium of technologists and environmentalists, or a progressive confrontation? These matters are of genuine concern to governmental agencies and legislative bodies around the world.

For those who make the decisions about how our planet is managed, there is an ongoing need for continual surveillance and intelligent controls to avoid endangering the environment, public health, and wildlife. Ensuring safety-in-use of the many chemicals involved in our highly industrialized culture is a dynamic challenge, for the old, established materials are continually being displaced by newly developed molecules more acceptable to federal and state regulatory agencies, public health officials, and environmentalists.

Reviews publishes synoptic articles designed to treat the presence, fate, and, if possible, the safety of xenobiotics in any segment of the environment. These reviews can be either general or specific, but properly lie in the domains of analytical chemistry and its methodology, biochemistry, human and animal medicine, legislation, pharmacology, physiology, toxicology, and regulation. Certain affairs in food technology concerned specifically with pesticide and other food-additive problems may also be appropriate.

Because manuscripts are published in the order in which they are received in final form, it may seem that some important aspects have been neglected at times. However, these apparent omissions are recognized, and pertinent manuscripts are likely in preparation or planned. The field is so very large and the interests in it are so varied that the editor and the editorial board earnestly solicit authors and suggestions of underrepresented topics to make this international book series yet more useful and worthwhile.

Justification for the preparation of any review for this book series is that it deals with some aspect of the many real problems arising from the presence of foreign chemicals in our surroundings. Thus, manuscripts may encompass case studies from any country. Food additives, including pesticides, or their metabolites that may persist into human food and animal feeds are within this scope. Additionally, chemical contamination in any manner of air, water, soil, or plant or animal life is within these objectives and their purview.

Manuscripts are often contributed by invitation. However, nominations for new topics or topics in areas that are rapidly advancing are welcome. Preliminary communication with the editor is recommended before volunteered review manuscripts are submitted.

Summerfield, NC, USA David M. Whitacre

Contents

The University of California-Davis Methodology for Deriving Aquatic Life Pesticide Water Quality Criteria 1
Patti L. TenBrook, Amanda J. Palumbo, Tessa L. Fojut, Paul Hann, Joseph Karkoski, and Ronald S. Tjeerdema

Index 157

Contributors

Tessa L. Fojut Department of Environmental Toxicology, College of Agricultural and Environmental Sciences, University of California, Davis, CA 95616-8588, USA, tlfojut@ucdavis.edu

Paul Hann Central Valley Regional Water Quality Control Board, Rancho Cordova, CA 95670, USA, phann@waterboards.ca.gov

Joseph Karkoski Central Valley Regional Water Quality Control Board, Rancho Cordova, CA 95670, USA, jkarkoski@waterboards.ca.gov

Amanda J. Palumbo Department of Environmental Toxicology, College of Agricultural and Environmental Sciences, University of California, Davis, CA95616-8588, USA, amanda.palumbo@scionresearch.com

Patti L. TenBrook Department of Environmental Toxicology, College of Agricultural and Environmental Sciences, University of California, Davis, CA 95616-8588, USA; USEPA Region 9, San Francisco, CA 94105, USA, tenbrook.patti@epamail.epa.gov

Ronald S. Tjeerdema Department of Environmental Toxicology, College of Agricultural and Environmental Sciences, University of California, Davis, CA 95616-8588, USA, rstjeerdema@ucdavis.edu

The University of California-Davis Methodology for Deriving Aquatic Life Pesticide Water Quality Criteria

Patti L. TenBrook, Amanda J. Palumbo, Tessa L. Fojut, Paul Hann, Joseph Karkoski, and Ronald S. Tjeerdema

Contents

1	Introduction		2
	1.1	Definition of Numeric Criteria	3
	1.2	Goal and Level of Organization to Protect	5
2	Data for Criteria Generation		6
	2.1	Kinds of Data	6
	2.2	Filling Data Gaps with Estimation Techniques	15
	2.3	Data Sources and Literature Searches	17
	2.4	Data Summaries of Ecotoxicity Data	20
	2.5	Data Evaluation	20
	2.6	Data Quantity – Ecotoxicity	32
	2.7	Data Reduction	35
3	Criteria Calculation		37
	3.1	SSD Procedure	38
	3.2	AF Procedure	60
	3.3	Averaging Periods	69
	3.4	Allowable Frequency of Exceedance	71
4	Water Quality Effects		76
	4.1	Bioavailability	76
	4.2	Mixtures	80
	4.3	Other Water Quality Effects	87
5	Check Criteria Against Ecotoxicity Data		87
	5.1	Sensitive Species	88
	5.2	Ecosystem and Other Studies	88
	5.3	Threatened and Endangered Species	88
6	Partitioning to Other Environmental Compartments		89

T.L. Fojut (✉)
Department of Environmental Toxicology, College of Agricultural and Environmental Sciences, University of California, Davis, CA 95616-8588, USA
e-mail: tlfojut@ucdavis.edu

6.1	Bioaccumulation/Secondary Poisoning	89
6.2	Harmonization with Sediment and Air Criteria	92
7	Assumptions and Limitations	93
7.1	Assumptions, Limitations, and Uncertainties of the UCDM	93
7.2	Data Generation to Improve Criteria Derivation	94
8	Guideline Format	95
9	The UCD Methodology	95
9.1	Goals and Definitions	95
9.2	Data	96
9.3	Acute Criterion Derivation	105
9.4	Chronic Criterion Derivation	114
9.5	Incorporation of Water Quality Effects into Criteria Compliance	117
9.6	Checking Criteria Against Ecotoxicity Data	124
9.7	Partitioning to Other Environmental Compartments	127
9.8	Reviewing Assumptions and Limitations to Derived Criteria	133
9.9	Final Criteria Statement	134
10	Summary	134
Acknowledgments		136
Appendix: Acute Chlorpyrifos Data Collected for Criteria Derivation Using the UCDM Derivation		137
References		142

1 Introduction

A national water quality criteria methodology was established in the United States (US) in 1985 (US Environmental Protection Agency; USEPA 1985).[1] Since then, several other methods for establishing water quality criteria have been developed around the world, incorporating recent advances in the field of aquatic toxicology using a variety of different approaches. The authors of a recent review compared existing methodologies and summarized the differences between them (see tables 4 and 5 in TenBrook et al. 2009). TenBrook et al. (2009) observed that although methods from the USEPA provided a good basis for calculating criteria, many newer methodologies added valuable procedures that could improve criteria generation. Of particular concern were cases having small data sets, for which the USEPA (1985) methodology does not allow criteria calculation and provides little guidance. In this review, we elaborate on the review of methodologies by TenBrook et al. (2009) and we propose a new methodology that combines features derived from the existing methodologies that have been determined to generate the most flexible and robust criteria. This new methodology also incorporates results from recent research in aquatic ecotoxicology and environmental risk assessment and is hereafter referred to as the University of California-Davis Methodology (UCDM).

---

[1] Chemical names and CAS numbers of all chemicals referred to in this article are given in Table 26.

The development of the UCDM was the second part of a larger project undertaken by the Central Valley Regional Water Quality Control Board (CVRWQCB) of California and the University of California-Davis to (1) review and summarize the differences between existing aquatic life water quality criteria methodologies (TenBrook et al. 2009), (2) establish a new aquatic life water quality criteria methodology that elaborates on the Tenbrook et al. (2009) review, and (3) derive criteria for pesticides of concern using the new methodology. Because this project was a collaboration with the CVRWQCB, the protection goal (Section 1.2) of the UCDM is derived from California water policy. This project focused on the Sacramento and San Joaquin River watersheds of the California Central Valley, which ecosystem is referred to in several instances; however, the UCDM is generally appropriate for any freshwater ecosystem in the United States. Additionally, simple modifications could be made to adapt this method for saltwater criteria or to other geographic areas.

Criteria derivation is a process that can be divided into a number of steps, starting with data collection and ending with numeric criteria, designed, within the limits of data availability, to be protective of aquatic life. In this review, the prospective procedures that may be used to achieve each step are presented, evaluated, and those most suitable are selected for use in the UCDM. The procedures of the UCDM were selected to handle a variety of data sets, including those as small as one datum, and the use of different derivation procedures are recommended depending on the size and diversity of data sets. A complete description of the steps needed to derive aquatic criteria by the UCDM is presented in Section 9, in the form of a standard operating procedure. The limitations of these water quality criteria values are discussed qualitatively, and where possible, quantitatively. However, no categorization was made as to what the criteria values should be used for, as that decision lies in the realm of policy.

As an example of the application of the UCDM, a full set of chlorpyrifos data was collected, evaluated according to the UCDM, and utilized to derive chlorpyrifos water quality criteria. This chlorpyrifos data set and 11 other pesticide data sets from the USEPA (1980a–g, 1986b, 2003c, 2005a) were used to compare several criteria derivation models and to calculate pesticide-specific default values used in the UCDM (Section 3). The chlorpyrifos data set is available as supplemental information in the Appendix.

Table 1 is provided to define the many acronyms and abbreviations used throughout this review.

1.1 Definition of Numeric Criteria

Water quality criteria are referred to in the literature by different terms, and such criteria may be used for different purposes depending upon how they are derived. In this review, numeric criteria will be defined as science-based values that are intended to protect aquatic life from the adverse effects of pesticides, without consideration of

Table 1 List of acronyms and abbreviations used in this review

ACE	Acute-to-chronic estimation
AChE	Acetyl cholinesterase
ACR	Acute-to-chronic ratio
AF	Assessment factor
A/NZ	Australia/New Zealand
ANZECC	Australia and New Zealand Environment and Conservation Council
APHA	American Public Health Association
ARMCANZ	Agriculture and Resource Management Council of Australia and New Zealand
ASTM	American Society for Testing and Materials
BAF	Bioaccumulation factor
BCF	Bioconcentration factor
BMF	Biomagnification factor
BSAF	Biota-sediment accumulation factor
CAS	Chemical Abstract Service
CCME	Canadian Council of Ministers of the Environment
CDFG	California Department of Fish and Game
CDPR	California Department of Pesticide Regulation
CVRWQCB	Central Valley Regional Water Quality Control Board
DENR	Department of Environment and Natural Resources
DOC	Dissolved organic carbon
DOM	Dissolved organic matter
EEC	European Economic Community
ECB	European Chemicals Bureau
ECOTOX	Database (see Table 4)
EC_x	Concentration that affects $x\%$ of exposed organisms
EU	European Union
EXTOXNET	Extension Toxicology Network (database)
FACR	Final acute-to-chronic ratio
FAV	Final acute value
FCV	Final chronic value
FT	Flow-through test
GMAV	Genus mean acute value
HA	Humic acid
HPLC	High-performance liquid chromatography
IC_x	Concentration that inhibits a defined effect of $x\%$ of exposed organisms
ICE	Interspecies correlation estimation
IRED	Interim re-registration eligibility decision
IUPAC	International Union of Pure and Applied Chemistry
K_H	Henry's law constant
K_{OW}	Octanol–water partition coefficient
L	Less relevant or less reliable rating
LC_x	Concentration lethal to $x\%$ of exposed organisms
LL	Less relevant, less reliable rating
LN	Less relevant, not reliable rating
LOEC	Lowest-observed effect concentration
LOEL	Lowest-observed effect level
LR	Less relevant, reliable rating

Table 1 (continued)

MATC	Maximum acceptable toxicant concentration
MRID	Master Record Identification number (USEPA)
MSD	Minimum significant difference
N	Not relevant or not reliable rating
NA	Not applicable
NC	Not calculable
NOEC	No-observed effect concentration
NOEL	No-observed effect level
OC	Organic carbon
OECD	Organization for Economic Co-operation and Development
OPP	Office of Pesticides Programs (USEPA)
QSAR	Quantitative structure activity relationship
PAH	Polynuclear aromatic hydrocarbons
pK_a	Acid dissociation constant
PRC	Performance reference compound
R	Relevant or reliable rating
RED	Re-registration eligibility decision
RIVM	National Institute of Public Health and the Environment, Bilthoven, The Netherlands
RL	Relevant, less reliable rating
RN	Relevant, not reliable rating
RPF	Relative potency factor
RR	Relevant and reliable rating
S	Static test
SR	Static renewal test
SACR	Secondary acute-to-chronic ratio
SETAC	Society of Environmental Toxicology and Chemistry
SMCV	Species mean chronic value
SMAV	Species mean acute value
SPMD	Semi-permeable membrane device
SSD	Species sensitivity distribution
$t_{1/2}$	Degradation half-life
TCE	Time–concentration–effect analysis
TEF	Toxic equivalency factor
UCDM	University of California-Davis Methodology
UK	United Kingdom
US	United States
USEPA	United States Environmental Protection Agency
USFDA	US Food and Drug Administration

defined water body uses, societal values, economics, or other non-scientific considerations. The numeric criteria we present herein correspond to the USEPA definition of numeric criteria, and it is the derivation of this type of number that is the subject of this review.

1.2 Goal and Level of Organization to Protect

In a previous review, TenBrook et al. (2009) stated that the goal of most water quality criteria methodologies is to protect ecosystems, and to do so, many aim for protection at the species level. In TenBrook et al. (2009), it was pointed out that the disappearance of a single species could result in the unraveling of community structure from complex interactions among species; this suggests that ecosystems may not be fully protected if water quality criteria are derived by a method that does not have the goal of protecting all species. Other examples of protection goals would be to protect at the individual level, genus level, or the functional level in an ecosystem. California water policy was used as the basis for determining at which level of ecosystem organization the methods protection goal should be targeted. The narrative objective of the CVRWQCB is to maintain waters free of "toxic substances in concentrations that produce detrimental physiological responses in plant, animal, or aquatic life" (CVRWQCB 2004). Therefore, procedures proposed in the UCDM have the goal of protecting at the species level to fully protect natural ecosystems and meet the policy mandate. More explicitly, the goal of the UCDM is to extrapolate from pesticide toxicity data available for selected species to a concentration that should not produce detrimental physiological effects in aquatic life.

2 Data for Criteria Generation

Criteria derivation requires ecotoxicological effects data. Good criteria must be based on good quality data on organisms of adequate taxonomic diversity. The most reliable and most certain criteria are derived from the largest and best quality data sets. The UCDM includes guidance on what kinds of data should be collected, where to collect it, and how to evaluate its quality.

2.1 Kinds of Data

For thorough evaluation of pesticide effects, it is necessary to collect physical–chemical and ecotoxicity data. The key data requirements essential to the UCDM are presented in Table 2. Ecotoxicity data are used directly in criteria derivation; the goal is to collect high-quality data from as many taxonomic groups as possible because the risk of under- or over-protection decreases as sample size increases (Aldenberg and Slob 1993).

Because the UCDM is for derivation of criteria for use in the United States, only data for freshwater species from families with reproducing populations residing in North America are recommended for criteria derivation. Notwithstanding, the UCDM recommends that all available high-quality data should be collected to provide supporting information or for derivation of acute-to-chronic ratios (ACRs). We chose to consider geographic distribution at the family level, rather than at the species level, because the USEPA has demonstrated that interspecies toxicity correlations work well at the family level (Asfaw et al. 2003; see Section 2.2.2). In the remainder of this section, we provide specific details regarding the kinds of ecotoxicity data that are most useful for criteria derivation.

Table 2 Kinds of data that should be collected for criteria derivation

Category	Data
Physical–chemical	BAF (bioaccumulation factor)
	BCF (bioconcentration factor)
	BMF (biomagnification factor)
	CAS (chemical abstract service number)
	Chemical formula
	Density
	IUPAC name
	K_H (Henry's Law constant)
	Log K_d (solid–water partition coefficient)
	Log K_{DOC} (dissolved organic carbon–water partition coefficient)
	Log K_{OC} (organic carbon–water partition coefficient)
	Log K_{OW} (octanol–water partition coefficient)
	Melting point
	Molecular weight
	pK_a (acid dissociation constant)
	S (aqueous solubility)
	Structure
	$t_{1/2}$ (half-life), hydrolysis, photolysis, biotic degradation
	Vapor pressure
Ecotoxicity	Acute (survival, immobilization)
	Aquatic insects
	Aquatic plants
	Bioavailability
	Chemical mixtures
	Chronic (survival, growth, reproduction, embryonic/shell development, hatching, germination, behavior effects, enzyme inhibition, endocrine disruption, other physiological effects, insect control, changes in species diversity or abundance)
	Field
	Fish
	Insects
	Laboratory
	Mesocosm
	Microcosm
	Multi-species
	Non-insect aquatic invertebrates
	Single chemical
	Single-species
	Wildlife
Human health	USFDA action levels

2.1.1 Acute vs. Chronic Toxicity Data

Both acute and chronic toxicity data are needed for criteria derivation. As discussed in TenBrook et al. (2009), the difference between acute and chronic toxicity is not always clear; therefore, it is important that we define in the UCDM which kinds of tests are considered to be acute vs. chronic. We use the USEPA (2003a) general definitions of acute and chronic:

> Acute toxicity: Concurrent and delayed adverse effect(s) that results from an acute exposure and occurs within any short observation period which begins when the exposure begins, may extend beyond the exposure period, and usually does not constitute a substantial portion of the life span of the organism.

> Chronic toxicity: Concurrent and delayed adverse effect(s) that occurs only as a result of a chronic exposure. Chronic exposure is exposure of an organism for any long period or for a substantial portion of its life span.

These general definitions are helpful, but for clarity more specific guidance is needed. Following are definitions of acute and chronic data from existing methodologies. Most of these are incorporated into the UCDM:

Acute

(1) Crustacean or insect tests lasting 24–96 h (Netherlands' National Institute for Public Health and the Environment [RIVM] 2001; Siepmann and Finlayson 2000; USEPA 1985, 2003a);
(2) Fish, mollusk, or amphibian tests lasting 96 h (RIVM 2001);
(3) Shellfish embryo, larval, or older life-stage tests lasting 96 h (USEPA 1985, 2003a).

Chronic

(1) Algae, bacteria, or protozoa tests lasting 3–4 days (RIVM 2001);
(2) Fish, mollusk, or amphibian early life-stage tests and 28-days growth tests (RIVM 2001);
(3) Single-celled organism tests of any duration (USEPA 1985, 2003a);
(4) Any test that takes into account the number of young produced, regardless of duration (USEPA 2003a);
(5) Full life cycle (ranging from 7 days for mysids to 15 months for salmonids), partial life cycle (all major life stages exposed for less than 15 months; specifically for fish that require more than 1 year to reach sexual maturity), and early life-stage tests (ranging from 28 to 60 days; also specifically for fish) (USEPA 1985, 2003a).

Endpoints in acute tests may be survival or immobility and endpoints in chronic tests may be survival, growth, reproduction, or measures of population growth rate. Other endpoints that have been linked to survival, growth, or reproduction may also be included. See Section 2.1.3 for further discussion of endpoints.

Definitions 1 and 3 in the chronic list apply to single-celled organisms. Similarly, definitions 2 and 4 in the chronic list apply to early life-stage, or short-term chronic tests. The USEPA definitions (3 and 4 above in the chronic list) are included in the UCDM because they will result in the inclusion of a broader range of tests than those from the RIVM document. For example, algae, bacteria, and protozoa tests shorter than 3 days are included by the USEPA definition, but excluded by the RIVM definition. Life cycles of plants vary widely, and procedures for conducting toxicity tests with plants are not well developed. Currently plant toxicity tests usually measure endpoints associated with chronic toxicity, such as growth and reproduction. Therefore, explicit definitions for acute plant tests are not included, and all plant and algal toxicity data will be considered chronic toxicity data.

Typically, very few chronic-study data are available for a given chemical. Methods are available for deriving chronic toxicity values and chronic water quality criteria from acute toxicity data. These methods are discussed in Sections 2.2.3 and 3.2.5.

2.1.2 Hypothesis Tests vs. Regression Analysis

Ecotoxicity test data are usually analyzed by one of two methods. Hypothesis tests, which are typically used for life-cycle, partial life-cycle, and early life-stage tests are designed to compare treatment groups to a control group for purposes of determining which treatment groups significantly differ from the control group (Stephan and Rogers 1985). A no-observed effect concentration (NOEC) or no-observed effect level (NOEL) and a lowest-observed effect concentration or level (LOEC or LOEL, respectively) may be derived from such testing. Also, some methodologies employ the geometric mean of the NOEC and LOEC to calculate a maximum acceptable toxicant concentration (MATC). The other widely used method for analysis of ecotoxicity data is regression analysis, which is most commonly used for acute toxicity tests, but can as easily be applied to chronic tests. In regression analysis, an equation is derived that describes the relationship between concentrations and effects (Stephan and Rogers 1985). Thus, it is possible to make point estimates of toxicant concentrations that will cause an effect (EC_x) to x percent of organisms, or to predict effects for a given level of toxicant. The effect may be lethality, expressed as (LC_x). Acute toxicity test results are most often expressed as LC_{50} values (the concentration that is lethal to 50% of tested organisms).

Many problems are associated with hypothesis testing, and these are described in the literature. They are summed up succinctly by Stephan and Rogers (1985), who point out that the effect value obtained from a hypothesis test is dependent on what toxicant concentrations were actually tested and the selection of α (type I error rate), which is usually arbitrarily chosen at 0.05. Hence, Hoekstra and Van Ewijk (1993) have shown examples in which NOEC test values can be misinterpreted as actual no-effect levels. In contrast, regression analysis determines a relationship between concentration and effect, and so provides a means to estimate (interpolate) the exposure at which untested concentrations may produce effects. Moreover, with regression analysis, the confidence limits will change according to α or with an increase in variability, but the point estimate will not change. Bruce and Versteeg (1992) observe another shortcoming of hypothesis testing; namely, when results are reported solely as a NOEC value, information on the concentration–response curve and variability in the data are lost.

There is apparent agreement among toxicologists that regression analysis provides better effect level estimates than do hypothesis tests (Bruce and Versteeg 1992; Grothe et al. 1996; Moore and Caux 1997; Stephan and Rogers 1985). Regression methods are commonly used and widely accepted for use in analyzing acute toxicity data, but for analysis of chronic data, hypothesis tests have been more widely used. Most chronic toxicity studies reported in the literature use hypothesis tests to determine the NOEC and the LOEC, making this the most available form of chronic data. In addition, although regression methods are preferred, there is little agreement among scientists as to what statistical exposure level should be considered to have no biological effect (e.g., EC_5, EC_{10}). The challenge, then, is to find a way to ensure that toxicity values derived from hypothesis tests are reliable values for use in criteria derivation.

A good approach to determine test reliability was proposed by participants in a 1994 workshop in The Netherlands (Van Der Hoeven et al. 1997). They concluded that NOEC data may be used as a summary statistic in ecotoxicity testing if the following are reported: (a) the minimum significant difference (MSD); (b) the actual observed difference from control; (c) the statistical test used; and (d) the test concentrations. These factors can be used to make case-by-case judgments as to whether a reported NOEC is reliable. For example, an extremely high MSD can be a sign of a poorly conducted test, in which high variability within treatment groups has obscured differences between groups. Thus, the MSD should be reported as a measure of within-test variability for NOEC data (Denton et al. 2003; USEPA 2002a). Similarly, the response at the NOEC should be reasonable compared to the control group (e.g., a 50% reduction compared to the control group should certainly not be considered "no effect"), the statistical methods should be appropriate, and the test should be designed with an appropriate dilution factor (test-specific, but ranging from 1.5 to 3.2 in various authoritative methods published by the Organization for Economic Co-operation and Development (OECD); the American Public Health Association (APHA); and the American Society for Testing and Materials (ASTM)).

To meet the standards of the UCDM, acute toxicity data should be in the form of LC_{50} or EC_{50} values derived from regression analysis. The use of chronic data expressed as EC_x values (from regression analysis) is only acceptable if species-specific studies are available to show what level of x represents a biological no-effect level. Species-specific studies are also required to determine what levels of effects can be detected in toxicity tests (Denton and Norberg-King 1996).

Chronic data expressed as results of hypothesis tests are acceptable, but must be evaluated to ensure that reported toxicity values are reasonable estimates of no-effect levels. The UCDM includes the following factors in test reliability rating schemes for results of hypothesis tests: MSD, observed difference from control at the NOEC, LOEC, or MATC, the statistical method used, and test concentrations. Absence of these parameters, or unacceptable results for them, will not necessarily eliminate a test from the data set, but will reduce its reliability score.

The question remains as to which hypothesis test value (NOEC, LOEC, MATC) should be used for criteria derivation. In a well-designed, well-conducted toxicity test, the true no-effect level lies between the NOEC and LOEC (each of which must be one of the toxicant concentrations actually tested). Although interpolation between these toxicity values is not strictly allowed (Stephan and Rogers 1985), the MATC, which is the geometric mean of the LOEC and NOEC, represents an accepted way of estimating the true no-effect level. Therefore, the MATC is the value used in the UCDM to calculate the chronic criterion.

2.1.3 Endpoints

As discussed in TenBrook et al. (2009), most derivation methodologies use ecotoxicity data only from studies in which the evaluated endpoints were survival and/or immobility, and growth and/or reproduction. Reproductive endpoints include histopathological effects on reproductive organs, spermatogenesis, fertility, pregnancy rate, number of eggs produced, egg fertility, and hatchability (RIVM 2001). These so-called traditional endpoints are favored because they can be readily linked to population-level effects. Linkages between effects at successive levels of organization are what define whether or not an observed effect is biologically significant (Suter and Barnthouse 1993). Non-traditional endpoints, such as endocrine disruption, enzyme induction, enzyme inhibition, behavioral effects (other than immobility), histological effects, stress protein induction, changes in RNA or DNA levels, mutagenicity, and carcinogenicity, have had very few links established between these effects seen at the individual level and effects at the population, community, or ecosystem level. Many of these non-traditional endpoints are merely markers of exposure, with no link between that exposure and adverse effects on survival, growth, or reproduction. Generally, non-traditional endpoints can be used as supporting information, but not directly in criteria derivation.

Two exceptions to this generalization are the endocrine disrupting effects of tributyltin and the inhibition of acetylcholinesterase (AChE) by pesticides. Segner (2005) discusses three cases in which population-level effects in wildlife could be linked to environmental substances with endocrine activity: reductions in dogwhelk (*Nucella lapillus*) populations due to imposex caused by exposure to tributyltin; reductions in predatory bird populations from egg-shell thinning caused by exposure to DDE; and a decline in Atlantic salmon populations from the effects of 4-nonylphenol on the ability of smolts to osmoregulate. However, only in the case of tributyltin is there a strong case for endocrine disruption as the mechanism of the observed toxic effects. As a result, the recent *Ambient Aquatic Life Water Quality Criteria for Tributyltin (TBT) – Final* (USEPA 2003b) utilizes data from several studies of imposex in gastropods to set the final chronic criterion.

The other exception refers to effects of AChE activity. Acute exposures to chlorpyrifos, at levels that caused mortality, have been linked to enzyme inhibition of >90% in larval walleye, and >71% in juvenile walleye (Phillips et al. 2002). Similarly, exposure to a variety of AChE-inhibiting chemicals, at levels that caused immobility, have been linked to enzyme inhibition of 50–70% in *Daphnia magna* (Barata et al. 2004; Printes and Callaghan 2004). On the other hand, Ferrari et al. (2004) found that goldfish exposed to azinphos-methyl could withstand cholinesterase reductions >90% without suffering mortality. Printes and Callaghan (2004) found that different cholinesterase-inhibiting pesticides had different inhibition levels associated with mortality. These studies indicate that although AChE activity has been associated with mortality, the association is species- and chemical-specific.

Whether AChE inhibition (or other biochemical endpoints) data can be used in criteria derivation has to be decided on a case-by-case basis. For example, in a study of Chinook salmon, Wheelock et al. (2005) determined a 96-h acetylcholinesterase inhibition NOEC for chlorpyrifos of 1.2 μg/L, LOEC of 7.3 μg/L, and MATC of 3.0 μg/L. At the LOEC, inhibition was 85% in brain tissue and 92% in muscle. Raw data from this study were analyzed by linear regression (Excel v. 11.2.5) to roughly estimate that 7.5% mortality would be expected at the MATC concentrations of 3.0 μg/L. Thus, in this case, a level of chlorpyrifos that causes less than 85% enzyme inhibition (i.e., at the MATC) is expected to result in 7.5% mortality. At the LOEC of 7.3 μg/L chlorpyrifos, which causes 85–92% enzyme inhibition, 18.2% mortality is expected. These levels of mortality would have to be shown to be both statistically and biologically significant for the AChE inhibition results to be used in criteria derivation. The statistical significance may be determined by analysis of toxicity test variability. For example, Denton and Norberg-King (1996) determined that a reduction of 11.2–14.9% in survival compared to controls could be detected in fish toxicity tests for some species. Biological significance is determined by a link to survival, growth, or reproduction because these are the endpoints recognized in the UCDM. In this example, AChE inhibition is linked to one of the recognized endpoints: survival. To be able to use AChE-inhibition, or other biochemical endpoints for criteria derivation in the UCDM, it would be necessary for authors of a toxicity study to derive an inhibition concentration (IC_x) value, where x is equal to the enzyme inhibition level that is linked to a statistically significant change in mortality, growth, or reproduction for a given chemical and species.

Another class of non-traditional endpoints is those that are directly linked to population-level effects, but are rarely determined in single-species toxicity studies. Population-level endpoints are suggested by Whitehouse et al. (2004) as a way to more directly predict toxic effects of chemicals on ecosystems. Using the population parameters r (intrinsic rate of population growth) and λ (factor by which a population increases in a given time), Whitehouse et al. (2004) found r to be a more sensitive endpoint than reproduction in tests with *D. magna* exposed to zinc. Moreover, they found EC_{20} values based on r to be in good agreement with a NOEC value determined for effects of 17α-ethinylestradiol in fathead minnows (Lange et al. 2001). On the other hand, Forbes and Calow (1999) found that, in most cases, r and λ were equally sensitive, or less sensitive to toxicant exposures compared to individual traits (e.g., reproduction). Whitehouse et al. (2004) note that while their intra-species examples of using population-level endpoints worked out well, there is little evidence to support their use across species for a given chemical. They attempted to collect data to expand the model across species, but out of 385 potentially usable studies, the needed mortality and fecundity data could only be obtained for six. Because of such lack of data, and lack of evidence that they are any more protective of ecosystems than traditional endpoints, population-level endpoints are not generally used in the UCDM, but can be used if available.

For the UCDM, results of tests using individual level endpoints – other than survival/immobility, growth, and reproduction – may be used to derive criteria if those endpoints have been adequately linked to effects on survival, growth, reproduction, or population-level parameters, such as r or λ. Population-level endpoints, such as r and λ, can be used if they come from studies rated as being relevant and reliable (RR), but only if a more sensitive endpoint is not available for that species.

2.1.4 Multispecies (Field/Semi-field, Laboratory) Data

Although there is much debate in the literature about whether or not single-species toxicity tests are good predictors of ecosystem effects, multispecies data are problematic for use in criteria derivation because of their paucity and variability. Several studies have shown that the repeatability, reproducibility, and ecological realism of these mesocosms were poor enough to preclude the use of such data in predictive risk assessment, or for extrapolation to natural ecosystems (Crane 1997; Hanson et al. 2003; Kraufvelin 1999; Sanderson 2002). As discussed by TenBrook et al. (2009), water quality criteria derived from single-species tests are protective of ecosystems in many cases. Moreover, it has been demonstrated in several studies that laboratory-derived NOECs were predictive of field effects (Borthwick et al. 1985; Crane et al. 1999; Persoone and Janssen 1994).

Because of these problems, and the relative cost-effectiveness, reproducibility, and reliability of single-species toxicity tests, most methodologies do not utilize multispecies or field data for criteria derivation. Some methodologies do not use field or semi-field data directly, but do use them as a comparison to criteria derived from single-species data (OECD 1995a; RIVM 2001). In some cases, a final criterion may be adjusted if strong multispecies evidence indicates that the single-species criterion is over- or under-protective (RIVM 2001; USEPA 1985, 2003c; Zabel and Cole 1999).

In view of the foregoing, multispecies data are not used for criteria derivation in the UCDM. Although not useful for direct derivation of criteria, field or semi-field data are very useful for comparison to criteria derived from single-species data (OECD 1995a; RIVM 2001), and may provide justification for downward adjustment of a final criterion (RIVM 2001; USEPA 1985, 2003a; Zabel and Cole 1999). If toxicity values, obtained for appropriate endpoints (i.e., those related to survival, growth, or reproduction) in high-quality multispecies studies, are lower than the derived criteria, then criteria may need to be adjusted downward.

Adjustment of criteria upward is not recommended because single-species data have indicated this concentration to be protective and raising the criterion may result in toxicity to sensitive species.

2.1.5 Data from Multi-pathway Exposures

Until food web or other models are further developed to incorporate multi-pathway exposures into criteria derivation, the best approach is to do water-only assessments (TenBrook et al. 2009). If studies show these criteria to be underprotective, and if the substance has a log-normalized octanol–water partition coefficient (log K_{OW}) values between 5 and 7, then dietary uptake studies, specific to the compound and species affected, should be performed to determine if exposure has been significantly underestimated. Water-only exposures are used for criteria derivation by the UCDM, but if derived criteria for chemicals with log K_{OW} values between 5 and 7 appear to be underprotective, then targeted studies are recommended to determine more precisely if dietary uptake is significant for particular species/chemical combinations. For pesticides with log K_{OW} values between 5 and 7, laboratory toxicity test data should be carefully reviewed to ensure that feeding regimes eliminated, or minimized any effects from interaction of the pesticide with food particles (e.g., reduction of test solution concentration as a result of partitioning into the food particles, or introduction of a dietary exposure route if animals ingest food that has had a chance to sorb pesticide).

2.1.6 Toxicity Data that Incorporate Time

To address the effect of time on toxicity of chemicals, it would be ideal to be able to derive criteria defined for any given exposure duration, thus eliminating the need to define separate acute and chronic criteria. In other words, different criteria would be derived for different exposure scenarios. Thus, environmental managers would be able to determine compliance with water quality standards that range from brief pulse exposures to infinite exposures. This could be done by, for example, segregating acute toxicity data according to exposure duration and deriving criteria for 24-, 48-, 72-h, etc., durations. Alternatively, toxicity test data could be analyzed using time-to-event methods to describe dose–time–response surfaces (Sun et al. 1995), or by kinetic models that describe the time course of toxicity (USEPA 2005b). TenBrook et al. (2009) observed that these models are data and resource intensive. Toxicity data, as they are currently (typically) reported, do not include information about multiple time points or kinetics. Thus, these time-to-event models are not currently feasible for general use in criteria derivation. Therefore, derivation of separate acute and chronic criteria is the approach adopted for the UCDM. However, as discussed in Section 2.2.3, if data are available, time–concentration–effect (TCE) models may be used to estimate chronic toxicity. Also if site- or situation-specific (i.e., for pulse exposures) criteria are desired, studies could be designed and conducted specifically to gather data for those models and the resulting toxicity data could be used to derive time-specific criteria.

2.2 Filling Data Gaps with Estimation Techniques

A common challenge in derivation of water quality criteria is that very few usable data are available. This is of particular concern in the case of chronic toxicity data. In this section we present approaches for estimation of acute and chronic toxicity. Some estimating techniques, such as ACRs, are widely used and accepted. Some, such as quantitative structure activity relationships (QSARs) are widely accepted for some kinds of toxicants, but are still under development for most toxicants. Others, such as TCE models are newer, have been validated for a large number of fish species, but require very data intensive procedures that are not feasible with most currently available data.

2.2.1 Quantitative Structure Activity Relationships (QSARs)

QSARs can be useful for estimation of acute and chronic toxicity for chemicals with non-specific modes of action (narcotic), but for reactive and specifically acting chemicals, QSAR models are not reliable for estimation of toxicity (TenBrook et al. 2009). The ANZECC and ARMCANZ (2000) guidelines caution that QSARs should not be used as black boxes and considerable chemical expertise is required in their use. Using QSARs only to predict the toxicity of narcotic chemicals minimizes the risk of inappropriate use. Many industrial chemicals (e.g., solvents or other hydrocarbons) have a narcotic mode of action, but most pesticides have specific modes of action. Therefore, QSARs are of limited use for pesticide criteria derivation until they are more fully developed for specifically acting chemicals.

For chemicals with non-specific modes of action, and for which no other ecotoxicity data exist, QSARs could be used to estimate toxicity, and those estimated toxicity values can be utilized in the species sensitivity distribution (SSD) and/or the assessment factor (AF) criteria derivation methods. However, if at least one measured toxicity datum is available, the AF procedure should be used as the basis to derive criteria (Section 3.2). Because the pesticides used in California must have been tested at least minimally for registration, there should not be cases in which measured toxicity data are unavailable. Therefore, the UCDM does not include a QSAR component as a way to supplement data sets. However, QSARs are included as an option for estimating toxicity to threatened and endangered species for narcotic-acting chemicals. The RIVM (2001) guidelines provide 19 QSARs for estimating chronic effects of narcotic chemicals on aquatic species representing nine different taxa. Other QSARs are available in the literature.

2.2.2 Interspecies Correlations for Estimation of Toxicity

The USEPA has developed interspecies correlation estimation (ICE) software (ICE v. 1.0), which can be used to estimate the acute toxicity of diverse compounds to aquatic species, genera, and families that have little or no measured toxicity data; such estimates are derived from species that have adequate data sets (Asfaw et al. 2003). Toxicity estimates made by interspecies correlations work well within taxonomic families, but less well as taxonomic distance increases. The ICE models generate estimated acute toxicity values that have confidence limits to quantify uncertainty. This technique is promising, but needs further development and validation before being used to generate data for criteria derivation. However, this model provides an excellent tool for assessment of potential effects of derived criteria on threatened and endangered species (discussed further in Section 5.3) and it is included for this use in the UCDM for that purpose.

2.2.3 Estimating Chronic Toxicity from Acute Data

Although acute toxicity data are often abundant, acceptable chronic data are often lacking, making it difficult to derive chronic criteria. Two approaches are available for estimating chronic toxicity from acute data. The first utilizes the ACR approach, and is readily workable, even with very small data sets. We elaborate on the ACR approach in Section 3.2.5, wherein we discuss the application of the technique during the criteria derivation process. The second approach is to use TCE models. These models are described in this section because they generate individual toxicity data that may be used in SSD derivation procedures.

The USEPA has developed an acute-to-chronic estimation program (ACE v. 2.0) that uses TCE models for estimating chronic toxicity from acute data (Ellersieck et al. 2003). Three slightly different models are included in the ACE v. 2.0 software package: the accelerated life testing model (Mayer et al. 2002; Sun et al. 1995), the multifactor probit analysis model (Lee et al. 1995; Mayer et al. 2002), and the linear regression analysis model (Mayer et al. 1994, 2002). Three important factors of toxicity are considered by each model: exposure concentration, degree of response, and time course of effect. Analysis of these three factors from acute toxicity test values allows for prediction of effects over long-term (chronic) exposures. The models only work for mortality data, thus will not provide estimates of chronic effects in cases where sub-lethal effects are more sensitive toxicity indicators. The Australia/New Zealand guidelines (ANZECC and ARMCANZ 2000) allow for the use of these same models for estimation of chronic toxicity from acute data.

Chronic values estimated by the ACE program models have been validated within species for seven fish species that were exposed to 17 different chemicals (Mayer et al. 2002), and across species for a variety of invertebrates and fish exposed to chlorpyrifos and ammonia (Whitehouse et al. 2004). Although TCE models hold promise for estimating chronic toxicity, lack of appropriate data presents a barrier to their use. Typically, acute toxicity data are reported as LC_{50} or EC_{50} values at one or more time points. However, the TCE models must be populated with response data for each concentration at multiple time points. This kind of information is normally collected in acute toxicity tests, but is rarely reported. Whitehouse et al. (2004) contacted authors of 385 toxicity studies to try to obtain the raw data needed for TCE modeling. Only 85 replies were received, and only 34 of those actually contained the needed data. When data are available, TCE models should be used to supplement chronic toxicity data sets. For the UCDM, the ACE program may be used to estimate chronic toxicity from acute data if appropriate data inputs are available. NOEC values derived from ACE may be used along with other chronic data in AF or SSD procedures.

2.3 Data Sources and Literature Searches

To ensure that toxicity and physical–chemical data are collected without bias, a thorough search of numerous sources is necessary. Although review articles and criteria documents are good starting points for data collection, original sources should be found and evaluated whenever possible. TenBrook et al. (2009) identified data sources from several existing criteria derivation methodologies. All of the sources mentioned are useful, and, as such, are simply compiled and presented, along with other useful sources in Tables 3 and 4 for the UCDM. The Dutch and Danish methodologies specify that literature searches should go back to 1970 and 1985, respectively (RIVM 2001; Samsoe-Petersen and Pedersen 1995). However, we recommend that any data available from the time a pesticide was first developed should be collected and evaluated.

Table 3 Data sources. Original sources identified through handbooks, review articles, etc., should be evaluated

Source	Details/Notes	Date(s)
US Environmental Protection Agency EPA re-registration eligibility decision (RED) or interim re-registration eligibility decision (IRED)	Review RED or IRED on compound and EPA Office of Pesticide Programs database (ipmcenters.org/Ecotox/). Submit FOIA request for relevant studies by completing an Affirmation of Non-Multinational Status form, available here: epa.gov/pesticides/foia/affirmation.htm, and sending with list of the study MRID numbers and info about yourself and who you work for, to: hq.foia@epa.gov	
California Department of Pesticide Regulation (CDPR)	Find relevant study numbers in the pesticide database: http://apps.cdpr.ca.gov/ereglib/ To retrieve studies, contact Registration Branch of CDPR: Jacquelyn Rivers: Jrivers@cdpr.ca.gov, or Rachel Kubiak: (916) 324-3939, rkubiak@cdpr.ca.gov.	

Table 3 (continued)

Source	Details/Notes	Date(s)
California Department of Fish and Game-Aquatic Toxicity Laboratory	Contact or check online for lab reports or criteria reports, may be available through CDPR	
University Libraries		
Electronic databases	See Table 4 for list and details	
Handbooks		
ECETOC	Aquatic toxicity data evaluation.	1993
Howard	Handbook of environmental fate and exposure data for organic chemicals. Vol. III: Pesticides	1991
Mackay et al.	Illustrated handbook of physical–chemical properties and environmental fate for organic chemical. Volume V. Pesticide chemicals	Book: 1997 CD-ROM: 1999
MITI	Biodegradation and bioaccumulation data on existing data based on the CSCL Japan	1992
Nikunen et al.	Environmental properties of chemicals	2003
Verschueren	Handbook of environmental data on organic chemicals, 5th edition	Print and CD_ROM: 2009
Others		
Review articles		
e.g., Racke	Environmental fate of chlorpyrifos	1993
e.g., Laskowski	Physical and chemical properties of pyrethroids	2002
Internal databases		
International criteria documents/ government reports	Often available via the internet	
Laboratory reports		
Manufacturer data	May be listed in RED/IRED, EPA OPP database and available from EPA, may be proprietary,	
Memos	May be listed in RED/IRED, EPA OPP database and available from EPA	
Registration packets	Studies used for pesticide registration may be listed in RED/IRED, EPA OPP database and available from EPA, packets can be difficult to obtain	

Table 4 Web addresses for various electronic resources

Database	Description/Contents	URL
CLOGP	K_{OW} calculator available through Bio-Loom	www.biobtye.com
BIOSIS	Bibliographic; multidisciplinary	http://www.biosis.org/
ChemFinder	Physical properties, chemical structures, and names	http://www.chemfinder.com
Chemical abstracts	Bibliographic; primarily chemistry, life sciences	http://www.cas.org/
Current contents	Bibliographic: multidisciplinary	http://scientific.thomson.com/products/ccc/
ECOTOX (was AQUIRE)	Single chemical toxicity information for aquatic and terrestrial life	http://www.epa.gov/ecotox/
EFDB	Environmental fate data base; access to DATALOG, BIOLOG, CHEMFATE, BIODEG	http://www.syrres.com/esc/efdb.htm
DATALOG	Bibliographic; environmental fate	
BIOLOG	Microbial toxicity and biodegradation	
CHEMFATE	Environmental fate and chemical-physical properties	
BIODEG	Biodegradation data	
EXTOXNET	Extension toxicology network; pesticide profiles and toxicology information	http://extoxnet.orst.edu/
Estimation Program Interface Suite	Tools from USEPA for estimation of numerous physical–chemical parameters	http://www.epa.gov/oppt; exposure/docs/episuite.htm
KowWin	Octanol–water partition coefficient program. Syracuse Research Corporation, New York, NY.	http://www.syrres.com/esc/est_soft.htm
LOGKOW	Sangster Research Laboratories	http://logkow.cisti.nrc.ca/logkow/index.jsp
Pesticide Action Network	Bibliographic; toxicity and regulatory information for pesticides	http://www.pesticideinfo.org/Index.html
PHYSPROP	Physical properties, chemical structures, and names	http://www.syrres.com/esc/physprop.htm
Pesticide Ecotoxicity Database	USEPA Office of Pesticide Programs toxicity database for registered pesticides, mostly unpublished studies	http://www.ipmcenters.org/Ecotox
POLTOX via OVID	Bibliographic; pollution and toxicology; plants, animals, and humans.	http://www.ovid.com
PubMed	Bibliographic; medicine, life sciences, molecular biology, genetics, others	http://www.ncbi.nlm.nih.gov/entrez/query.fcgi?DB=pubmed
TOXNET	Access to HSDB, TOXLINE, IRIS	http://toxnet.nlm.nih.gov/
HSDB	Hazardous Substances Data Bank	
TOXLINE	Toxicology Literature Online	
IRIS	Integrated Risk Information System	

Table 4 (continued)

Database	Description/Contents	URL
TSCATS	Bibliographic; Toxic Substances Control Act submission data	http://www.syrres.com/esc/tscats.htm
Web of Science	Bibliographic; access to Institute for Scientific Information Citation Databases	http://scientific.thomson.com/products/wos/

2.4 Data Summaries of Ecotoxicity Data

When reviewing ecotoxicity data, it is important to have a systematic way of rating the quality of a given study. To rate diverse studies fairly requires that similar information be obtained from each study. This particularly applies to single-species toxicity data; multispecies laboratory and field or semi-field tests are usually too complex to lend themselves to simplified summarization. To facilitate the data evaluation process in The Netherlands (RIVM 2001), data are put into data tables with the following headings: species (including scientific name), species properties (e.g., age, weight, life stage), analysis of test compound (measured or not, Y or N), test type (flow-through, static-renewal, static), substance purity, test water, pH, water properties (e.g., hardness, salinity), exposure time, test criterion (e.g., LC_{50} or NOEC), ecotoxicological endpoint (growth, reproduction, mortality, immobilization, morphological effects, histopathological effects), LC_{50} values, NOEC values, notes, and reference information.

Such a summary is helpful, but does not include enough information for thorough evaluation of the quality of the study. Critical information that is missing from the Dutch data summary tables includes the following: where test organisms reside, control response, source of test organisms with confirmation that they were collected from non-polluted sites, detailed test design information, detailed water quality information, concentrations of any carrier solvents used, statistical methods used, and whether responses recorded at NOEC and LOEC concentrations are reasonable. All of this information is needed to adequately rate the quality of test data, as defined in the methods outlined below, in Section 2.5. Therefore, a new, more detailed data summary table has been developed to use with the UCDM (Table 5). The table includes all of the elements mentioned above, as well as space to record any additional information that may be important to the study.

2.5 Data Evaluation

In this section, the issue of data quality is explored. High-quality data are regarded to be both relevant and reliable. The EU Technical Guidance Document on Risk Assessment describes reliability as the inherent quality of a test relating to test methodology and the way the performance and results of the test are described;

Table 5 Toxicity data summary sheet

Study:		
Relevance	Reliability	
Score:	Score:	
Rating:	Rating:	

Reference

Parameter	Value	Comment
Test method cited		
Phylum		
Class		
Order		
Family		
Genus		
Species		
Family in North America?		
Age/size at start of test/growth phase		
Source of organisms		
Have organisms been exposed to contaminants?		
Animals acclimated and disease-free?		
Animals randomized?		
Test vessels randomized?		
Test duration		
Data for multiple times?		
Effect 1		
Control response 1		
Effect 2		
Control response 2		
Effect 3		
Control response 3		
Temperature		
Test type		
Photoperiod/light intensity		
Dilution water		
pH		
Hardness		
Alkalinity		
Conductivity		
Dissolved Oxygen		
Feeding		
Purity of test substance		
Concentrations measured?		
Measured is what % of nominal?		
Chemical method documented?		
Concentration of carrier (if any) in test solutions		
Concentration 1 Nom/Meas (μg/L)		Reps and # per (cell density for single-celled organisms):
Concentration 2 Nom/Meas (μg/L)		Reps and # per (cell density for single

Table 5 (continued)

Concentration 3 Nom/Meas (μg/L)	Reps and # per (cell density for single
Concentration 4 Nom/Meas (μg/L)	Reps and # per (cell density for single
Concentration 5 Nom/Meas (μg/L)	Reps and # per (cell density for single
Control; describe type	Reps and # per (cell density for single
LCx; indicate calculation method	Method:
ECx; indicate calculation method	Method:
NOEC; indicate calculation method, significance level (*p* value) and minimum significant difference (MSD)	Method: p: MSD:
LOEC; indicate calculation method and significance level (*p* value)	Method: p:
MATC (GeoMean NOEC, LOEC)	
% control at NOEC	
% of control LOEC	

Other notes:

relevance refers to the extent to which a test is appropriate for a particular hazard or risk assessment (ECB 2003). Reliable data are obtained from studies in which test reports describe test details and indicate that tests were conducted according to generally accepted standards. Relevance is judged by whether appropriate endpoints were tested in a study, whether the study was conducted under relevant conditions, and if the substance tested was representative of the substance being assessed. Existing approaches for evaluation of physical–chemical and ecotoxicity data are discussed below, and the strongest elements of many are included in the UCDM.

2.5.1 Physical–Chemical Data

In the context of deriving water quality criteria, physical–chemical data are relevant to the extent that they enhance the interpretation of toxicity data. For example, the organic carbon (OC)–water partition coefficient (K_{OC}) can be used to predict concentrations of freely dissolved hydrophobic contaminants in water, and whether dietary uptake might be important. The acid dissociation constant (pK_a) can be used to predict the predominant form of an ionizable compound, and the half-life ($t_{1/2}$) can be helpful in determining if separate criteria are needed for degradation products. Toxicity tests can be evaluated based on whether test solution renewals were adequate, considering the $t_{1/2}$ of the compound, or whether reported toxicity values (i.e., LC_{50}, NOEC) are reasonable given the water solubility of the compound. K_{OW} values may be used, in limited cases, to predict the toxicity of chemicals using QSARs, if toxicity data are lacking. To the extent that physical–chemical parameters are affected by temperature, attention should be paid to whether a reported value was measured at a relevant temperature; if not, then physical–chemical values should be adjusted as necessary (RIVM 2001; Schwarzenbach et al. 1993).

TenBrook et al. (2009) concluded that reliable physical–chemical data are those determined by current, standard methods (e.g., ASTM, OECD, APHA) applied and performed correctly for the chemical of interest. Non-standard methods may also be appropriate, but only if valid reasons are given for deviation from standard methods, or if studies were done prior to the existence of standard methods, but generally followed currently acceptable practices. In regard to pesticides, which vary widely in chemical characteristics such as hydrophobicity, water solubility, and ionizability, it is particularly important to verify that reported partition coefficients were determined correctly. Thus, it is not acceptable to simply use a value reported in a handbook without verifying the value as it appeared in the original reference. An exception to this is that recommended K_{OW} values may be taken without further review from the LOGKOW database (Sangster Research Laboratories 2004) because all values in this database have been reviewed by partition coefficient experts.

Several web-based programs are available that can be used to estimate various physical–chemical parameters. These include ClogP3 (www.biobyte.com), the USEPA's Estimation Program Interface Suite (www.epa.gov/oppt;exposure/docs/episuite.htm), and KowWin (www.syrres.com/esc/est_soft.htm). Several databases containing physical–chemical data are also available on the web, including the Environmental Fate Database (www.syrres.com/esc/efdb.htm), EXTOXNET (extoxnet.orts.edu), PHYSPROP (www.syrres.com/esc/physprop.htm), TOXNET (toxntet.nlm.nih.gov), and the US Department of Agriculture, Agricultural Research Service Pesticide Properties Database (http://www.ars.usda.gov/services/docs.htm?docid=14199). These are helpful resources, but should be used with caution. Estimated parameters should only be used in the absence of measured data. As with physical–chemical values taken from handbooks, any values taken from online databases should be fully referenced as to their original sources and those sources should be checked to ensure that the values were appropriately obtained.

Although it is preferable to use only physical–chemical data from studies that can be reviewed and verified, in practice sometimes, the only information available may be in a handbook that cites unverifiable data. Such values should be used with caution. In some cases, values can be considered reliable without further review. For example, recommended values from the LOGKOW database (Sangster Research Laboratories 2004), because of the careful review process to which the values in this database were subjected. Another example is the physical–chemical data from unpublished manufacturer studies that normally have wide acceptance. If several values are available for the same physical–chemical parameter, and all were obtained by acceptable methods, the geometric mean of the values measured at the same temperature should be used.

In the UCDM, we provide general guidance on the acceptability of physical–chemical data based on information in several existing methodologies (OECD 1995a; RIVM 2001; USEPA 1985, 2003a). More specifically, tables are provided that specify acceptable experimental and computational test methods for determining physical–chemical parameters such as log K_{OW}, water solubility, dissociation constants (pK_a), and hydrolysis or other degradation rates

Table 6 Acceptable methods for determination of physical–chemical parameters, other than the octanol–water partition coefficient, K_{OW}

Constant	Method	Notes	Reference
Bioconcentration Factor, BCF	Flow-through; fish	Determines apparent steady-state BCF	OECD 305 (1996)
	Flow-through; fish and mollusks	Determines apparent steady-state BCF	ASTM E 1022-94 (2002a)
Dissociation, pK_a	Conductometric	Onsager (1927) equation must hold; Acid/base dissociations; Non-acid/base dissociations	OECD 112 (1981)
	Spectrophotometric	Solubility: low to high; Differential UV/vis absorption for ionized vs. unionized species; Acid/base dissociations; Non-acid/base dissociations	"
	Titration	Solubility: moderate to high	"
Hydrolysis Rate	Tiered approach	Determines rate in acidic, basic and neutral conditions	ASTM E895-89 (2001a)
	Tiered approach	Determines rate in acidic, basic and neutral conditions	OECD 111 (2004)
Solid–water partition, K_d, K_{OC}	Batch Equilibrium	Colloidal binding can reduce accuracy	ASTM E 1195-01 (2001b)
	Batch Equilibrium	Colloidal binding can reduce accuracy	OECD 106 (2000)
	Batch Equilibrium Co-solvent	Corrects for colloid binding	Evers and Smedes (1993)
	HPLC	Estimation technique	OECD 121 (2001)
Solubility, S	Column Elution	Solubility $<10^{-2}$ g/L	OECD 105 (1995b)
	Flask	Solubility $>10^{-2}$ g/L	"
	Flask	Solubility ≥ 1 mg/L	ASTM E 1148-02 (2002b)
	Generator column	Solubility <1 mg/L	"
	Nephelometric	Solubility ≥ 1 mg/L	"

(Tables 6 and 7). Moreover, several on-line databases and calculators that may be used to obtain or estimate physical–chemical parameters are provided in Table 4.

Table 7 Acceptable experimental and computational techniques for determination of the octanol–water partition coefficient, K_{OW}, and the priority for their use (USEPA 2003a)

Log K_{OW} <4

Method	Reference	Priority
Slow stir	Debruijn et al. (1989)	1
Generator-column	USEPA (1996a)	1
Shake-flask	USEPA (1996b)	1
HPLC w/ extrapolation to 0% solvent	ASTM E 1147-92 (1997)	2
HPLC w/o extrapolation to 0% solvent	ASTM E 1147-92 (1997)	3
CLOGP program	Through Bio-Loom at www.biobtye.com	4

Log K_{OW} >4

Method	Reference	Priority
Slow stir	Debruijn et al. (1989)	1
Generator-column	USEPA (1996a)	1
HPLC w/ extrapolation to 0% solvent	ASTM E 1147-92 (1997)	2
HPLC w/o extrapolation to 0% solvent	ASTM E 1147-92 (1997)	3
Shake-flask	USEPA (1996b)	4
CLOGP program	Through Bio-Loom at www.biobtye.com	5

2.5.2 Ecotoxicity Data Evaluation

By studying criteria derivation methodologies (TenBrook et al. 2009) and working through the process of collecting and evaluating chlorpyrifos effects data, it became clear that ecotoxicity data must be evaluated on three points: (1) relevance to criteria derivation; (2) documentation; and (3) acceptability. Documentation and acceptability together define the reliability of a study. The ECOTOX (2006) system for rating documentation of aquatic and terrestrial toxicity data from laboratory and field is widely accepted and is included in the UCDM, with some modifications by way of added detail. A similar system is used for evaluation of acceptability and relevance. The elements upon which a test is judged are similar for rating documentation and acceptability, but in the former case, scores are based solely upon whether or not an item was documented, whereas in the latter case, scores are based upon whether or not a given parameter was within accepted testing guidelines and organism tolerances. Weighting of scores for acceptability is based upon test acceptability criteria, as stated in standard methods (e.g., USEPA, OECD, APHA, or ASTM toxicity test methods). For example, control response and temperature control, which are common measures of test acceptability, are weighted more heavily as measures of reliability than water hardness and alkalinity, which are not as critical. As there are no standard methods for performing multispecies lab, field or mesocosm/microcosm, or wildlife studies, these studies should be rated primarily on documentation, but also on a few key acceptability criteria as described by OECD (1995a) and RIVM (2001). The UCDM includes detailed data scoring schemes, with point values assigned to each toxicity test element (Tables 8, 9, 10, 11, and 12).

Table 8 Rating of relevance/usability of single-species data for derivation of criteria

Parameter	Score
Acceptable standard (or equivalent) method used	10
Endpoint linked to survival/growth/reproduction	15
Freshwater	15
Chemical ≥80% pure	15
Species is in a family that resides in North America	15
Toxicity value calculated or calculable (e.g., LC_{50})	15
Controls	15
Described (i.e., solvent, dilution water)	7.5
Response reported and meets acceptability requirements	7.5
Total	100

The elements for judging relevance of a study are somewhat different, but can be weighted and rated in a similar fashion. The relevance scores should be weighted such that failure of just one of a number of critical factors immediately renders the study irrelevant for criteria derivation. Tests that involve in vitro exposures of organs or tissues (i.e., were not whole-body exposures) and tests in which toxicity values >2x of the water solubility of the pesticide are reported are not useful even as supporting information and can be eliminated without further consideration. For compounds with log K_{OW} between 5 and 7, laboratory test feeding regimes should eliminate or minimize interaction of pesticide with food particles (see Section 2.1.5). All other effect studies should be evaluated for relevance based on the following critical factors: controls must be documented and must meet minimum test acceptability requirements; tests must be performed on species belonging to families that reside in North America; endpoints must be linked to survival, growth, or reproductive effects; tests must produce toxicity values (i.e., no > or < values); tests must be with freshwater species; and tests must be conducted with pesticide samples that are ≥80% pure (i.e., no formulations and no mixtures). Only single-species toxicity tests need to be evaluated for relevance because they are the only ones that may be used directly for criteria derivation. Tests that fail to meet any one of these criteria may not be used to derive criteria or for derivation of ACRs, but may be used as supporting information provided they are rated highly enough.

The decision to include ecotoxicity data, when reported toxicity values are up to 2x greater than the water solubility of the pesticide, is based on a recent review by Shen and Wania (2005), as well as on data from the PAN and PHYSPROP databases (PAN 2006; PHYSPROP 2006). These sources show that water solubility values

Table 9 Documentation rating for single-species aquatic laboratory data (adapted from ECOTOX 2006). Full score is given if parameter is reported; 0 score is given if not

Parameter[a]	Score[b]
Results published or in signed, dated format	6
Exposure duration	12
Control type	8
Organism information (i.e., age, life stage)	
Source	5
Age/life stage/size/growth phase	5
Chemical	
Grade or purity	5
Analytical method (if measured)	4
Nominal concentrations	3
Measured concentrations	3
Exposure type	5
Dilution water source	3
Hardness	2
Alkalinity	2
Dissolved oxygen	4
Temperature	4
Conductivity	2
pH	3
Photoperiod and/or light intensity (plant studies must include intensity)	3
Statistics	
Methods identified	5
Hypothesis tests	
Statistical significance	2
Significance level	2
Minimum significant difference	2
% of control at NOEC and/or LOEC	2
Point estimates (i.e., LC_{50}, EC_{25})	8
Total	100

[a]Compiled from RIVM (2001), USEPA (1985, 2003a), ECOTOX (2006), CCME (1999), ANZECC and ARMCANZ (2000), OECD (1995a), and Van Der Hoeven et al. (1997).
[b]Weighting based acceptability criteria from various ASTM, OECD, APHA, and USEPA methods, ECOTOX (2006), and on data quality criteria in RIVM (2001), USEPA (1985, 2003a), CCME (1999), ANZECC and ARMCANZ (2000), OECD (1995a), and Van Der Hoeven et al. (1997).

reported for a given pesticide can vary by greater than 100-fold, but it is common for reported water solubility values to vary by a factor of 2. Thus it seems reasonable to accept toxicity data that are within a factor of two of the geometric mean of acceptable solubility values.

For single-species toxicity data in this evaluation scheme, documentation and acceptability scores are averaged, resulting in one score for reliability. The reliability and relevance scores are then used to give the final rating. The scores for the two categories (relevance and reliability) are set (see Table 13) so that studies with the

Table 10 Acceptability rating for single-species aquatic laboratory data (adapted from ECOTOX 2006). Score is given if parameter met standard test guidance; score of 0 is given if parameter was not reported or did not meet test guidance

Parameter[a]	Score[b]
Acceptable standard (or equivalent) method used (e.g., ASTM, USEPA, OECD, APHA)	5
Test was of appropriate duration	2
Control	
Appropriate (e.g., solvent control included, if carrier was used)	6
Response within test guidance	9
Chemical	
Purity >80% pure	10
Measured concentrations within 20% of nominal	4
Concentrations do not exceed 2× water solubility	4
Carrier solvent ≤0.5 mL/L (acute); ≤0.1 mL/L (chronic); score 4 if not used	4
Organisms	
Appropriate size/age/growth phase	3
No prior contaminant exposure	4
Organisms randomly assigned to test containers	1
Adequate number per replicate/appropriate cell density	2
Organisms fed 2 h before solution renewal or not fed in acute tests; fed appropriately in chronic tests	3
Organisms properly acclimated and disease-free prior to testing	1
Exposure type and renewal frequency appropriate to chemical	2
Dilution water source acceptable	2
Hardness within organism tolerance and/or dilution water specifications	2
Alkalinity within organism tolerance and/or dilution water specifications	2
Dissolved oxygen ≥60%	6
Temperature within organism tolerance (3 pts) and/or test guidance and held to ± 1°C (3 pts)	6
Conductivity within organism tolerance and/or dilution water specifications	1
pH within organism tolerance and/or dilution water specifications	2
Photoperiod and light intensity within organism tolerance and/or test guidance	2
Statistics	
Adequate number of concentrations	3
Random or random block design employed	2
Adequate replication	2
Appropriate spacing between concentrations (dilution factor ≥0.3)	2
Appropriate statistical method used	2
Hypothesis tests	
Minimum significant difference (MSD) below recommended upper bound[c]	1
NOEC response reasonable compared to control[d]	1
LOEC response reasonable compared to control[d]	1

Table 10 (continued)

Parameter[a]	Score[b]
Point estimates	
LC/EC values calculable (i.e., no < or > results)	3
Total	100

[a]Compiled from RIVM (2001), USEPA (1985, 2003a), ECOTOX (2006), CCME (1999), ANZECC and ARMCANZ (2000), OECD (1995a), and Van Der Hoeven et al. (1997).
[b]Weighting based acceptability criteria from various ASTM, OECD, APHA, and USEPA methods, ECOTOX (2006), and on data quality criteria in RIVM (2001), USEPA (1985, 2003b), CCME (1999), ANZECC and ARMCANZ (2000), OECD (1995a), and Van Der Hoeven et al. (1997).
[c]Acceptable MSD levels are species and test-method specific; see USEPA (2002b) for upper bounds for several standard test species.
[d]Reasonableness is decided using professional judgment on a case-by-case basis, based on MSD upper bound and potential biological significance of response level.

highest scores are considered relevant/reliable (*R*), studies receiving middle scores are considered less relevant/less reliable (*L*), and the lowest scoring studies are considered not relevant/not reliable (*N*). When the two categories are combined, the final ratings fall into one of three categories: (1) those that may be used for criteria derivation (rating = RR); (2) those that may be used as supporting information (rating = RL, LR, LL); and (3) those that are not usable (any with an *N* in the rating).

To establish a rating scale, the chlorpyrifos data set collected for this report, given in the Appendix, was evaluated using the scoring systems detailed in Tables 8, 9, and 10. The chlorpyrifos set was broken down as follows: scores in the 75th or higher percentile of all scores were rated reliable; scores between the median and the 75th percentile were rated less reliable; and scores below the median were rated unreliable. Similarly, relevance scores in the 90th or higher percentile were rated relevant; scores between the median and the 90th percentile were rated less relevant; scores below the median were rated not relevant. The 75th percentile of scores is suggested for the reliability rating because, in the case of the chlorpyrifos data set, higher percentiles were too restrictive, resulting in rejection of too much data that others have accepted for criteria derivation (Siepmann and Finlayson 2000; USEPA 1986a). On the other hand, the selection of the 75th percentile resulted in rejection of a few tests accepted by others, indicating that this rating system is a bit more rigorous than those used previously. The relevance scoring system was designed to include six major requirements for a study to be used in criteria derivation. Lack of one of these requirements would lower the score below 90, so only studies scoring above 90 should be used for criteria derivation. Based on this translation of numeric scores into ratings for the chlorpyrifos data set, the numeric scale presented in Table 13 was established for rating other pesticide data sets in the UCDM.

All other chlorpyrifos effects data (i.e., not single-species) were evaluated on documentation and reliability, using the scoring schemes in Tables 11 and 12.

Table 11 Documentation and acceptability rating for aquatic outdoor field data and indoor model ecosystems (adapted from ECOTOX 2006)

Parameter[a]	Score[b]
Results published or in signed, dated format	5
Exposure duration and sample regime adequately described	6
Unimpacted site (score 7 for artificial systems)	7
Adequate range of organisms in system (1° producers, 1°, 2° consumers)	6
Chemical	
Grade or purity stated	6
Concentrations measured and reported	2
Analysis method stated	2
Habitat described (e.g., pond, lake, ditch, artificial, lentic, lotic)	6
Water quality	
Source identified	3
Hardness reported	2
Alkalinity reported	2
Dissolved oxygen reported	2
Temperature reported	2
Conductivity reported	2
pH reported	2
Photoperiod reported	2
Organic carbon reported	2
Chemical fate reported	3
Geographic location identified (score 2 for indoor systems)	2
Pesticide application	
Type reported (e.g., spray, dilutor, injection)	2
Frequency reported	2
Date/season reported (score 2 for indoor systems)	2
Test endpoints	
Species abundance reported	3
Species diversity reported	3
Biomass reported	2
Ecosystem recovery reported	2
Statistics	
Methods identified	2
At least 2 replicates	3
At least 2 test concentrations and 1 control	3
Dose response observed	2
Hypothesis tests	
NOEC determined	4
Significance level stated	2
Minimum significant difference reported	2
% of control at NOEC and/or LOEC reported or calculable	2
Total	100

[a]Compiled from RIVM (2001), USEPA (1985, 2003a), ECOTOX (2006), CCME (1999), ANZECC and ARMCANZ (2000), OECD (1995a), and Van Der Hoeven et al. (1997).
[b]Weighting based ECOTOX (2006) and on data quality criteria in RIVM (2001) and OECD (1995a).

Table 12 Documentation and acceptability rating for terrestrial laboratory/field data (adapted from ECOTOX 2006). Score is given if parameter is reported

Parameter[a]	Score[b]
Exposure duration	20
Control type	7
Organism information (i.e., age, life stage)	8
Chemical grade or purity	5
Chemical analysis method	5
Exposure type (i.e., dermal, dietary, gavage)	10
Test location (i.e., laboratory, field, natural artificial)	5
Application frequency	5
Organism source	5
Organism number and/or sample number	5
Dose number	5
Statistics	
Hypothesis tests	
Statistical significance	5
Significance level	5
Minimum significant difference	3
% of control at NOEC and/or LOEC	3
Point estimates (i.e., LC_{50}, EC_{25})	4
Total	100

[a] Compiled from ECOTOX (2006) and Van Der Hoeven et al. (1997).
[b] Weighting based on ECOTOX (2006).

Table 13 Data categories based on relevance and reliability scores. N = not relevant/not reliable; L = less relevant/reliable; R = relevant, reliable. The unshaded category (RR) is used for criteria derivation; the RL, LL, and LR categories are used for supporting data; the RN, LN, NN, NL, and NR categories are not usable

		Reliability		
	Score	0–59	60–73	74–100
	0–69	NN	NL	NR
Relevance	70–89	LN	LL	LR
	90–100	RN	RL	RR

Aquatic and terrestrial studies receiving *R* or *L* ratings may be used as supporting data in the criteria review process; studies rated *N* may not be used.

The data evaluation system presented above has been incorporated into the UCDM. The system includes procedures for scoring relevance and reliability of single-species, multispecies, laboratory, field, semi-field, microcosm, and mesocosm studies, as well as a table describing ranges of scores that define the categories RR, RL, RN, etc.

2.6 Data Quantity – Ecotoxicity

The UCDM includes specific instructions regarding the number and taxonomic diversity of data required for criteria derivation by SSD and AF procedures. First discussed is the number of data required.

Several current methodologies use toxicity data from five species to derive criteria by statistical extrapolation. The Australian/New Zealand methodology (ANZECC and ARMCANZ 2000) uses five single-species chronic NOEC values. OECD (1995a) guidelines use statistical extrapolations by Aldenberg and Slob (1993) or Wagner and Løkke (1991) that require at least five chronic NOEC values. The California Department of Fish and Game (CDFG) has derived criteria using the USEPA (1985) SSD procedure that requires fewer than the eight families; in this procedure, professional judgment is used to determine if species in the missing categories were relatively insensitive and their addition would not lower the criteria (Menconi and Beckman 1996; Siepmann and Jones 1998).

Although more data would improve fit, the use of five data values for statistical extrapolations by parametric techniques has been supported in the literature (Aldenberg and Luttik 2002; Okkerman et al. 1991). According to Aldenberg and Slob (1993) the risk of underprotection of a 50% confidence limit estimate of the hazardous concentration for 5% of organisms (HC_5; based on a log-logistic distribution) decreases considerably as sample size is increased from 2 to 5, but less so as it is increased from 5 to 10 and from 10 to 20. Considering the chosen distribution (described in Section 3.1), the Burr Type III distribution and software, BurrliOZ v. 1.0.13, is used with a minimum of five data values (ANZECC and ARMCANZ 2000). Therefore, five data values are the minimum required for criteria derivation by the UCDM.

When five ecotoxicity data values are unavailable (data sets as small as one datum), only AF derivation methods are appropriate (see Sections 3 and 3.2 for description of AFs). The minimal data sets available for use in derivation of criteria in California will be those required for registration under the Federal Insecticide, Fungicide, and Rodenticide Act (FIFRA; US Code Title 7 1947), and those required by the California Department of Pesticide Regulation (CDPR; CDPR 2005). According to 40 CFR Part 158.490 (USEPA 1993) the minimum data required for federal US registration under FIFRA is an LC_{50} for a fish and an LC_{50} for a freshwater invertebrate. All other kinds of aquatic toxicity data are only conditionally required, depending on planned pesticide usage, the potential for transport to water, whether any acute LC/EC_{50} values were <1 mg/L, whether estimated environmental concentrations are >0.01 times any LC/EC_{50}, or if data indicate reproductive toxicity, persistence, or bioaccumulative potential. It is possible that for many new chemicals, only the two acute toxicity data values will be available. The CDPR has tiered data requirements (CDPR 2005). The minimum data set includes LC_{50} values for one warm water fish, one cold water fish, and for a freshwater invertebrate. Further testing is required for the same reasons described in FIFRA. Again, it is possible that no more than the minimum data will be available for criteria derivation for new pesticides. An AF criteria derivation method is needed for these very small data sets.

In addition to data quantity, it is also important to specify the range of taxonomic diversity that should be represented in the data set. The USEPA methodologies (USEPA 1985, 2003a) have the most comprehensive, specific taxonomic requirements among the reviewed methodologies and are used for the UCDM, with some modification to reflect species of importance to the Central Valley of California and to reduce the required number of data values from eight to five.

First, consider the following taxonomic requirements in the USEPA (1985, 2003a) methods:

The following are required for derivation of acute or chronic criteria by the SSD procedure (minimum of eight acute or chronic data values):

(a) the family Salmonidae in the class Osteichthyes;
(b) one other family (preferably a commercially or recreationally important warm water species) in the class Osteichthyes (e.g., bluegill, channel catfish);
(c) a third family in the phylum Chordata (e.g., fish, amphibian);
(d) a planktonic crustacean (e.g., cladoceran, copepod);
(e) a benthic crustacean (e.g., ostracod, isopod, amphipod, crayfish);
(f) an insect (e.g., mayfly, dragonfly, damselfly, stonefly, caddisfly, mosquito, midge);
(g) a family in a phylum other than Arthropoda or Chordata (e.g., Rotifera, Annelida, Mollusca);
(h) a family in any order of insect or any phylum not already represented.

With regard to plants, the USEPA (1985) guidelines require results from at least one test that utilizes a freshwater alga or vascular plant, whereas the Great Lakes methodology (USEPA 2003a) indicates that plant data are desirable, but not required. In both cases, if plants are among the most sensitive (as is likely with herbicides), then tests with plants representing at least two phyla are required.

For derivation of acute criteria by the AF procedure:

(a) At least one datum from the family Daphniidae; species must be from the genus *Daphnia*, *Ceriodaphnia*, or *Simocephalus* (USEPA 2003a).

For derivation of an ACR (minimum of three chronic data values):

(a) at least one fish;
(b) at least one invertebrate;
(c) at least one acutely sensitive freshwater species (the other two may be saltwater species).

With these requirements in mind, and with the minimal data sets available from pesticide registration, the following taxonomic requirements are in the UCDM:

For derivation of acute or chronic criteria by the SSD procedure (minimum of five data values):

(a) the family Salmonidae;
(b) a warm water fish;
(c) a planktonic crustacean, of which one must be in the family Daphniidae in the genus *Ceriodaphnia, Daphnia,* or *Simocephalus*;
(d) a benthic crustacean;
(e) an insect (aquatic exposure).

The rationale for exclusion of items c, g, and h in USEPA list is as follows:

(c) two fish species (one warm water and one cold water) are sufficient to represent the phylum Chordata;
(g) rotifers, annelids, and mollusks are typically insensitive to pesticides (e.g., Giesy et al. 1999);
(h) this category is very general and simply fills out the eight minimum data values required by the USEPA SSD procedure (USEPA 1985, 2003a).

For determination of acute criteria by the AF procedure (minimum of one datum):

(a) The family Daphniidae in the genus *Ceriodaphnia, Daphnia,* or *Simocephalus*.
(b) Additional data must be from different families as per the list of those required for the SSD procedure. For example, to derive insecticide criteria, if data are available from two acceptable studies, then one must be from the family Daphniidae and the other must be either a member of the family Salmonidae, or a warm water fish, or a benthic crustacean or an insect. If three data values are available, then one must be in the family Daphniidae and the others must be from two other, different families, and so on, such that each additional datum contributes toward completion of the minimum data set required for the SSD procedure. This is to ensure that the magnitude of the AF is only reduced in cases where data are available for multiple families and to encourage generation of data that would complete the minimum SSD set.

For determination of ACRs, the requirements in USEPA guidance (USEPA 1985, 2003a) are acceptable, including the use of saltwater species if not enough freshwater data are available.

Additionally, alga or vascular aquatic plant data must be included for herbicides. The plant requirement for herbicides is added because herbicides are expected to be more toxic to plants than to animals. Because life cycles of plants vary widely, procedures for conducting toxicity tests with plants are not well developed, and explicit definitions for acute plant tests are not included, the methodology for herbicides will differ slightly.

Plant data are not required for the acute distribution, but an acute criterion should still be derived with animal data according to the requirements above. If the chemical is an herbicide and plants are the most sensitive group, options for chronic data are as follows:

(1) Distribution: Fit a distribution with only alga or vascular aquatic plant data, if there are data from at least five different species that were rated RR;
(2) Non-distributional: If data values from the five different required species are unavailable, or a distribution cannot be fit, use the lowest NOEC value from an important alga or vascular aquatic plant species that has measured concentrations, providing the endpoint is biologically relevant.

The non-distributional option for deriving a chronic criterion of an herbicide is similar to the USEPA Final Plant Value from their 1985 methodology. An alternative method is needed for herbicides because without acute alga or plant data an ACR cannot be calculated for plants. Additionally, the AF procedure included in the UCDM would not be appropriate as these factors were derived for acute data based on animal requirements; however, AF may still be used to derive an acute value for herbicides based on animal data.

Many pesticides are fungicides. The mode of action for fungicides, however, is not specific to fungi, so fish or invertebrates may be very sensitive to the toxic effects of fungicides. As for herbicides, ecological risk assessment for fungicides is a topic that has not received much attention. Maltby et al. (2009) found that using plant and animal data in the SSD appeared to derive protective criteria; however, there are very limited data on fungi to assess effects on non-target fungi. The new method will require the same five taxa for animal taxa described above for fungicides.

2.7 Data Reduction

Data that have been rated RR require further reduction prior to criteria derivation so that no species receives undue weight in the derivation process. For example, if there are multiple data for a particular species, then some method has to be specified for reducing those data into a single point within the SSD for that species. As discussed in TenBrook et al. (2009) the geometric mean is a reasonable approach for reducing data from multiple tests to a single number for criteria derivation. No species should be represented more than once in the final SSD.

SSD procedures assume that toxicity data in the distribution represent an unbiased sample of the system to be protected (Forbes and Calow 2002). Current USEPA procedures (USEPA 1985, 2003a) utilize genus mean toxicity values on the grounds that this reduces potential bias as a result of the overabundance of data for species from a few genera, and it minimizes statistical problems that arise because of non-random sampling (i.e., the close relationship between organisms within a genus prevents organisms from responding independently). This approach does not seem entirely justified because there is considerable variability of sensitivity between species within a genus in some cases. For example, Harmon et al. (2003) reported an EC_{50} (immobility) for *Daphnia ambigua* exposed to chlorpyrifos of 0.035 μg/L, whereas Van Der Hoeven and Gerritsen (1997) reported an EC_{50} (immobility) for

Daphnia pulex of 0.42 µg/L, a tenfold difference. In addition, sets of toxicity data acceptable for criteria derivation are usually quite small (e.g., in Erickson and Stephan 1988, data sets ranged from 8 to 45 values); thus, general statements about the intra-genus variability are not well-supported. For these reasons, and since no other existing methodology uses the genus-level approach, the UCDM utilizes data at the species level for both acute and chronic criteria derivation.

Specific data reduction procedures used in existing methodologies were described by TenBrook et al. (2009). Following is a compilation of those methods that are used for reducing data to species mean acute values (SMAVs) and species mean chronic values (SMCVs) in the UCDM.

(1) Calculate SMAVs/SMCVs as the geometric mean of toxicity values from one or more acceptable tests with the same endpoints (ANZECC and ARMCANZ 2000; ECB 2003; OECD 1995a; RIVM 2001; USEPA 1985, 2003a).
(2) If data are available for life stages that are at least a factor of two more resistant than another life stage for the same species, then do not use the data for the more resistant life stage to calculate the SMAV because the goal is to protect all life stages (RIVM 2001; USEPA 1985, 2003a).
(3) If data are available for one species, but for multiple endpoints, then use the data for the most sensitive endpoint (ANZECC and ARMCANZ 2000; ECB 2003; OECD 1995a; RIVM 2001); if multiple endpoints are equally sensitive, then note both endpoints, but use only one value for criteria calculation.
(4) If a NOEC is not explicitly reported in chronic toxicity studies, but statistical analysis was done, the NOEC may be determined as the highest reported concentration not statistically different from the control ($p < 0.05$; RIVM 2001); the NOEC is not used in criteria derivation, but is needed for calculation of the MATC.
(5) Similarly, if a LOEC is not explicitly reported in chronic toxicity studies, it may be determined as the lowest reported concentration that is statistically different from the control ($p < 0.05$); the LOEC is not used in criteria derivation, but is needed for calculation of the MATC.
(6) If a MATC is not reported, it may be calculated as the geometric mean of the NOEC and LOEC.
(7) If no toxicity values were reported, but raw data are available, calculate toxicity values using appropriate statistical methods (ECB 2003).
(8) If a MATC is expressed as a range of values, recalculate the MATC as the geometric mean of the high and low values (RIVM 2001).
(9) If reasons for differences between tests for the same species/endpoints are found, then data may be grouped according to appropriate factors (e.g., pH or temperature; ECB 2003). Selection of the appropriate value to use in criteria derivation should be based on standard test parameters. Tests conducted under non-standard conditions (vs. standard conditions as defined in standard test methods) may be used to derive quantitative relationships between those conditions and toxicity (as in USEPA 1985, 2003a). If such a relationship is established then toxicity values derived under non-standard conditions may be

translated to standard conditions and added to the criteria derivation data set. If no quantitative relationship can be derived then tests conducted under non-standard conditions should not be used for criteria derivation, but may be used as supporting information.

(10) If data are available for multiple time points from crustacean or insect acute toxicity studies, use the latest time point (i.e., 96-h tests are preferred over tests of <96-h duration).

(11) For a given species, use data from flow-through tests in which concentrations were measured, if available. If such data are not available, then data from static or static-renewal tests and/or tests in which concentrations were not measured may be used as long as they are rated otherwise relevant and reliable.

(12) Further reduction may be needed in the course of SSD analysis. If data cannot be described by or fitted to a distribution, then the set should be examined for outliers and/or bimodality. The UCDM includes detailed guidance for detection of outliers and exclusion of outliers. If data are bimodally distributed (as determined visually), use only the lower of the two groups for criteria derivation (ANZECC and ARMCANZ 2000); the effects of data exclusions on the criteria must be explored and explained (ECB 2003).

Data reduction procedures excluded from this list include those that equate NOECs with some percentage reduction from control responses and those that use factors to estimate NOECs from LOECs, with factor size dependent on level of response (ECB 2003; RIVM 2001). As discussed in Section 2.1.2, there is little agreement as to what levels of effect constitute "no effect" (in terms of statistical or biological significance), so these extrapolations are unreliable. The data reduction procedure of the UCDM does not include a USEPA procedure to exclude data for species in cases where toxicity values differ by more than a factor of 10 (USEPA 1985, 2003a). This particular approach for excluding data is not explained (USEPA 1985, 2003a) and step 12 in the data reduction procedure should adequately manage outliers. Also excluded was a procedure whereby multiple data for a given species were reduced by consideration of which studies best reflected regional environmental parameters (ECB 2003). This procedure makes little sense for laboratory toxicity data obtained in standardized tests, but could be useful in choosing field or semi-field data to support the criteria derivation process.

Based on this discussion, guidance is provided in the UCDM for selecting or calculating values for use in criteria derivation. Instructions are given for how to reduce multiple data for a given chemical/species combination to a single SMAV or SMCV, and for how to manage bimodal distributions and outliers.

3 Criteria Calculation

Criteria need to have components of magnitude, duration, and frequency to be most useful to environmental managers (TenBrook et al. 2009). Exposure duration is partially addressed through the derivation of separate acute and chronic criteria

(USEPA 1985, 2003a). Further consideration of allowable duration and frequency of exceedances are discussed in Sections 3.3 and 3.4. The use of SSD- and AF-extrapolation methods for determination of criteria magnitudes are addressed in this section.

The aim of both SSD and AF procedures is to extrapolate from available toxicity data for a limited number of species to toxicity values that will be protective of all species in an ecosystem. The AF method involves multiplying the lowest value of a set of toxicity data by a factor to arrive at a criterion. The SSD procedure involves the use of one of several similar statistical extrapolation techniques (described in the next section) to determine the criterion. TenBrook et al. (2009) concluded that if at least five data values are available from five taxonomic groups (Section 2.6), then the SSD procedure should be used. If fewer than five data values are available, then the AF procedure should be used. There are a number of approaches to each of these methods, and the purpose of this section is to compare approaches using several example data sets, and to select the best one for inclusion in the UCDM.

3.1 SSD Procedure

The SSD procedure relies on a statistical probability distribution to determine the criterion. An SSD is used to model the variability of species sensitivities to a toxicant (i.e., toxicity data). Such distributions can be used to estimate concentrations that are likely to fall below the sensitivity of a portion of species (typically 95%); thus, use of SSD procedures aims to protect most species in the ecosystem. A general description of an SSD is presented by TenBrook et al. (2009).

There are five points on which SSD methodologies may differ: (1) the shape of the distribution; (2) the kinds and quantity of data; (3) the level of confidence associated with derived criteria; (4) how data are aggregated; and (5) what percentile cutoff is the best predictor of no-effect concentrations. Item 2 has been addressed in Sections 2.1 and 2.6. Items 1, 3, and 4 will be further addressed below. Item 5 was discussed in TenBrook et al. (2009); the 5th percentile was determined to be the best predictor of no-effect levels. Because the percentile cutoff is such a critical issue in criteria derivation, we will briefly revisit the topic here.

3.1.1 Distribution Shape and Fit

Of the five above points, the shape of the distribution is one of the most important considerations in choosing a distribution. The shape of the distribution determines how well it fits the data. The SSD approach assumes that the toxicity data are a random sample of all species and that if all species were sampled, they could be described by some defined distribution. When a specific distribution is chosen, the assumption is that toxicity data from all species in an ecosystem (if it were obtainable) would fit the shape of that specific distribution. The goal is to try to select a distribution that fits the data sufficiently well to minimize violations of that distributional assumption.

In TenBrook et al. (2009), distributions used by major agencies around the world were reviewed for estimation of community or ecosystem effects that were based on single-species toxicity tests. In this section, the fit of the three most commonly used distributions are tested: the log-triangular, log-normal, and Burr Type III distributions. The goodness of fit and distributional assumptions of these three techniques were evaluated by testing the fit of the distribution to existing data sets. Twelve data sets were used: the chlorpyrifos acute toxicity data collected for this project (Appendix) and 11 acute pesticide data sets from USEPA criteria documents (USEPA 1980a–g, 1986a, b, 2003c, 2005a). These resultant data are shown in Table 14. Toxicity values in the USEPA diazinon and atrazine sets that were reported

Table 14 Acute toxicity data sets. Cpf = chlorpyrifos; Txp = toxaphene; End = endrin; Lin = lindane; Ald = aldrin; Diel = dieldren; Hept = heptachlor; Chl = chlordane; Endos = endosulfan; Dia = diazinon; Atr = atrazine; all toxicity values in μg/L

Rank	Cpf[a]	DDT[b]	Txp[b]	End[b]	Lin[b]	Ald[b]	Diel[b]	Hept[b]	Chl[b]	Endos[b]	Dia[b]	Atr[b]
1	0.035	0.36	0.8	0.15	2.0	4.0	2.5	0.9	3.0	0.34	0.3773	3,000
2	0.0427	1.1	1.3	0.32	10	4.5	4.5	1.1	6.3	0.83	0.7764	5,300
3	0.06	1.4	1.962	0.33	10.5	6.1	5.0	1.8	15	2.3	1.048	6,300
4	0.0654	1.6	2	0.41	32	7.4	6.1	2.8	26	3.2	1.587	6,700
5	0.100	1.7	2.3	0.44	32	8.0	8.0	7.8	26	3.7	2.04	14,700
6	0.150	1.7	3	0.46	40	9.0	8.1	13.1	37	3.8	6.51	20,000
7	0.220	1.9	3.1	0.47	44	10	10.8	23.6	40	5.8	10.7	27,000
8	0.25	1.9	3.446	0.54	44	16	15	24	45	6.0	16.82	49,000
9	1.0	2.4	3.7	0.69	45	21	20	26	56	88	25	60,000
10	4.7	2.6	3.822	0.75	48	27	22	29	57	261	425.8	
11	6	3.0	4.874	0.76	55.6	27	24	42	58		459.6	
12	8	3.0	5.782	0.78	64	28	39	47.3	59		602	
13	10	3.2	6	0.85	67.1	32	41	61.3	82		723	
14	15.96	3.5	6.7	1.0	68	34	130	78	190		800	
15	178	3.9	10	1.1	83	42	213	81.9			1,643	
16	806	4.0	10.12	1.2	90	45.9	250	101			2,166	
17	2410	4.3	10.8	1.3	138	50	567	148			3,198	
18		4.9	11.85	1.5	141.1	143	620	320			7,804	
19		5.0	12	1.8	207	180	740				7,841	
20		7.3	13	2.1	460						8,000	
21		7.8	13	3.1	485						9,000	
22		7.8	13.78	4.7	676						11,000	
23		8.0	14.59	5.9							11,640	
24		8.5	14.6	32								
25		9.3	15.68	34								
26		10	16.71	60								
27		12	17.61	64								
28		14	18	352								
29		17	20									
30		18	24									
31		25	26									
32		33	31.75									
33		40	40									
34		48	73.48									
35		48	140									
36		54	210									
37		67	500									
38		68										
39		175										
40		192										
41		362										

[a]Data collected for this project
[b]USEPA criteria documents (USEPA 1980a–g, 1986a, b, 2003c, 2005a)

as > or < values were excluded from this analysis because they are not usable in any but the log-triangular distribution method. The highest DDT value and the two highest aldrin toxicity values were excluded because these values each were greater than two times the water solubility of the respective compound (as discussed in Section 2.5.2).

Figure 1 (Parts 1–4) shows the log-normal distributional fits, along with box plots and histograms, of each data set (output from JMP IN v. 5.1.2, JMP 2004). Figure 2 (Parts 1–7) shows log-triangular, log-normal, and Burr Type III fits for these data sets (constructed using Excel v. 11.2.3). The log-triangular distributions were constructed according to the following parameters (Evans et al. 2000):

$$\text{CDF} = \frac{(x-a)^2}{(b-a)(c-a)} \text{ for } a \leqq x \leqq c; \tag{1}$$

$$\text{CDF} = \frac{(b-x)^2}{(b-a)(b-c)} \text{ for } c \leqq x \leqq b; \tag{2}$$

$$\text{Mean} = \frac{(a+b+c)}{3} \tag{3}$$

where

CDF = cumulative distribution function;
x = value in data set;
a = minimum value in data set;
b = maximum value in data set;
c = mode.

The mode of each log-triangular distribution was determined by dividing the data into bins and taking the average of the maximum and minimum values in largest bin.

The log-normal and Burr Type III distributions (Fig. 2, Parts 1–7) were constructed using Excel (v. 11.2.3). The Burr III distribution has the following cumulative distribution function:

$$F(x) = \frac{1}{\left[1 + \left(\frac{b}{x}\right)^c\right]^k} \tag{4}$$

where

$F(x)$ = probability of x;
b, c, and k are fit parameters;
x = data point in the distribution.

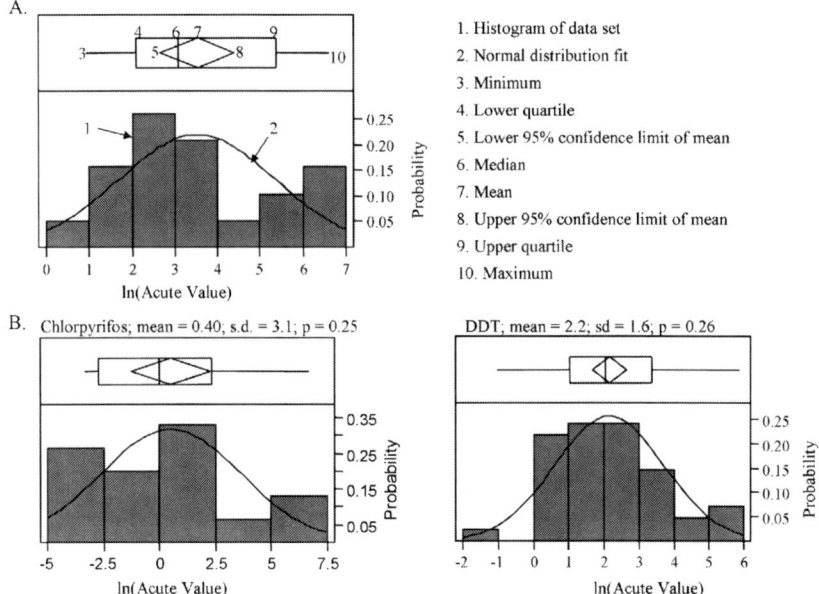

Fig. 1 (Part 1) Tests for log-normal distributions of data sets. A. Key to distribution diagrams. B. Distributions for chlorpyrifos and DDT data sets; $p < 0.05$ indicates lack of fit. ■ indicates outliers (outside 1.5 times the interquartile range)

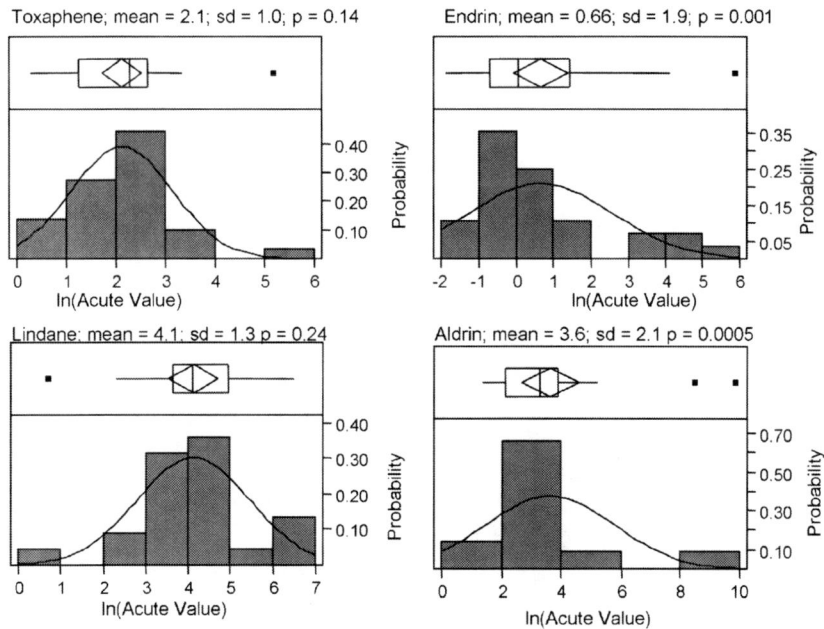

Fig. 1 (Part 2) Tests for log-normal distributions of data sets. Distributions for toxaphene, endrin, lindane, and aldrin data sets; $p < 0.05$ indicates lack of fit. ■ indicates outliers (outside 1.5 times the interquartile range)

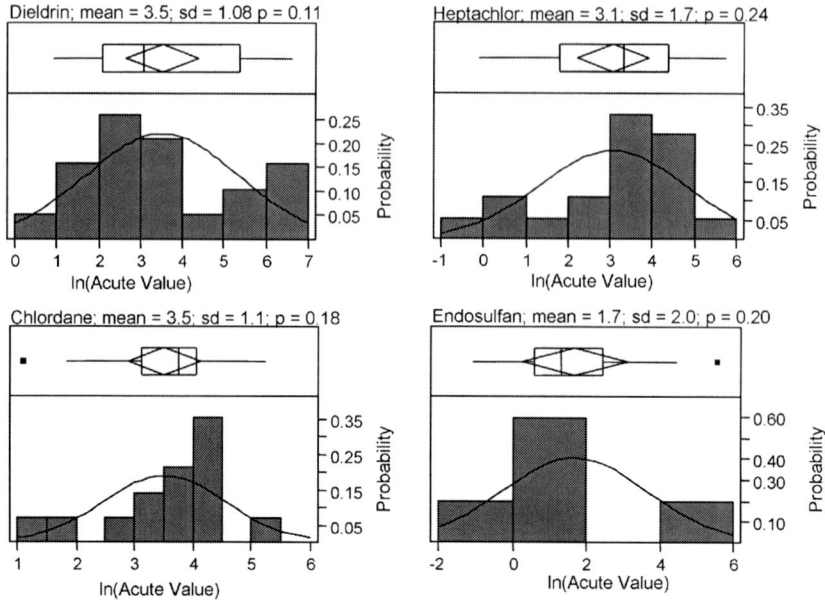

Fig. 1 (Part 3) Tests for log-normal distributions of data sets. Distributions for dieldrin, heptachlor, chlordane, and endosulfan data sets; $p < 0.05$ indicates lack of fit. ■ indicates outliers (outside 1.5 times the interquartile range)

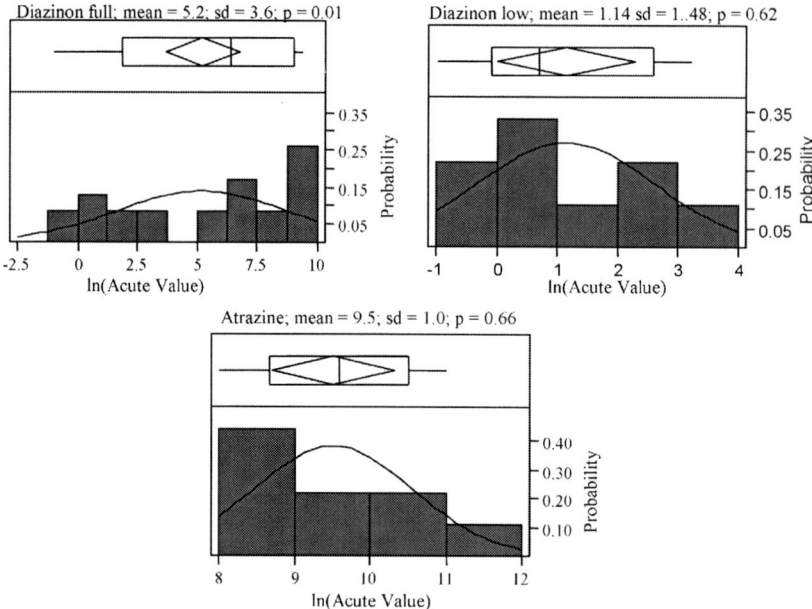

Fig. 1 (Part 4) Tests for log-normal distributions of data sets. Distributions for the full diazinon, lower subset of diazinon, and atrazine data sets; $p < 0.05$ indicates lack of fit. ■ indicates outliers (outside 1.5 times the interquartile range)

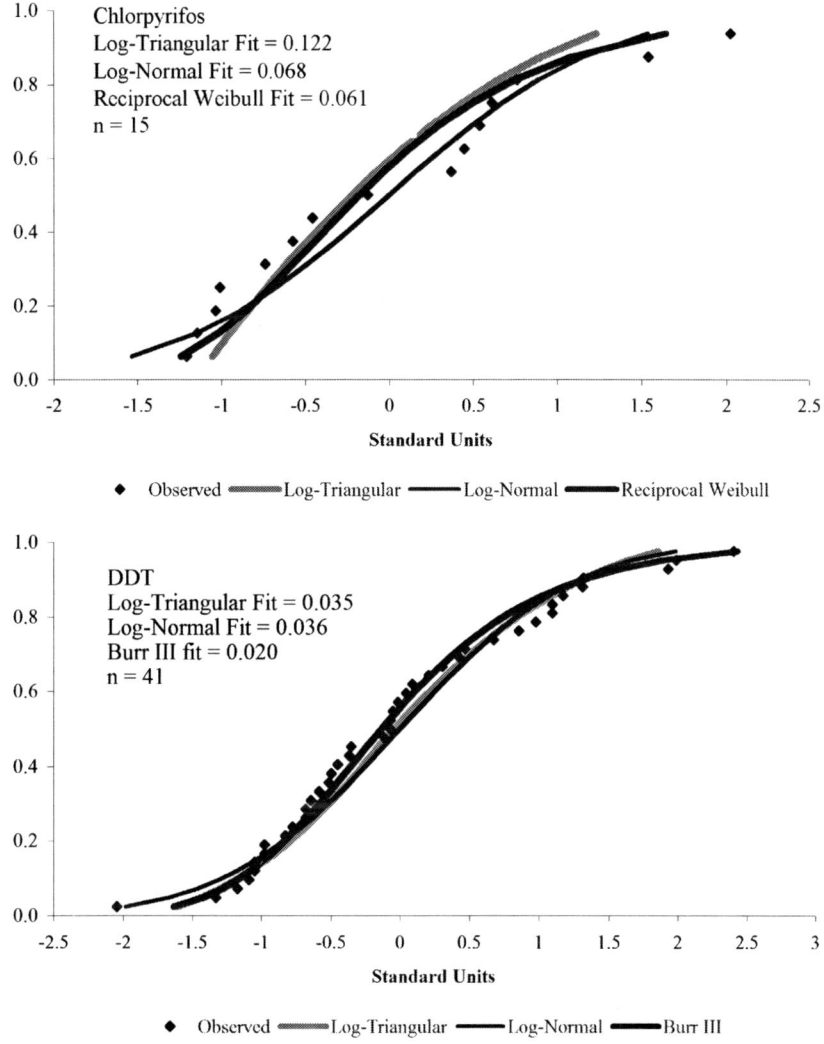

Fig. 2 (Part 1) Comparison of fits of pesticide toxicity data to log-triangular, log-normal, and Burr Type III distributions (including Reciprocal Weibull) for chlorpyrifos and DDT

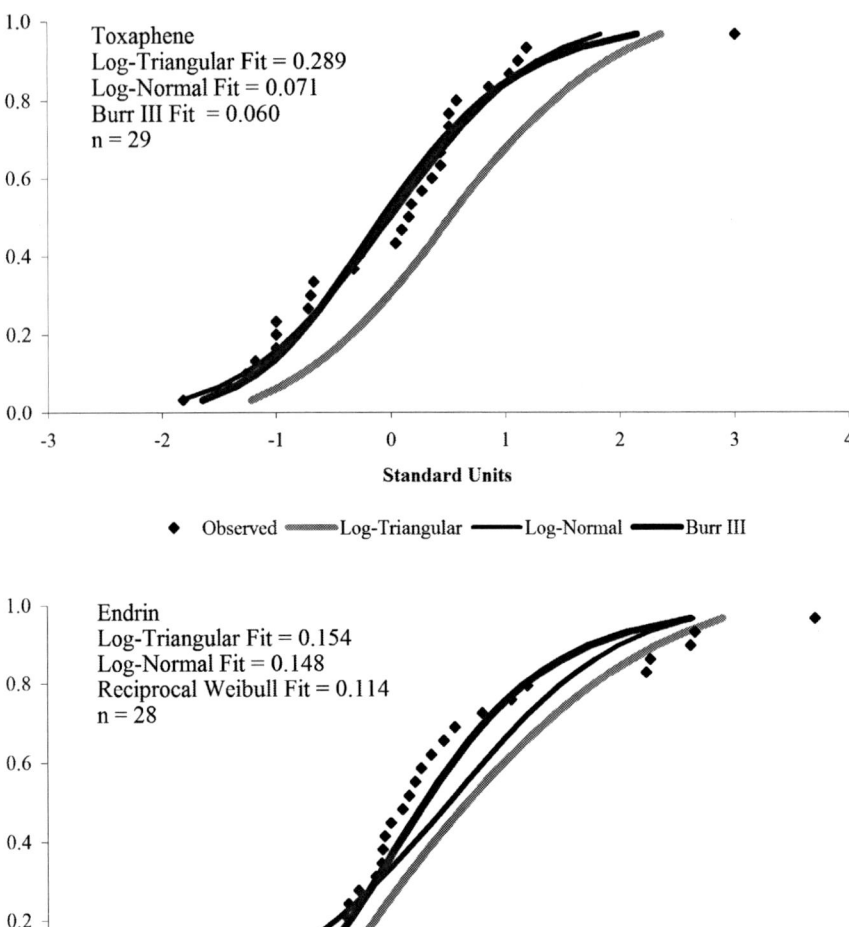

Fig. 2 (Part 2) Comparison of fits of pesticide toxicity data to log-triangular, log-normal, and Burr Type III distributions (including Reciprocal Weibull) for toxaphene and endrin

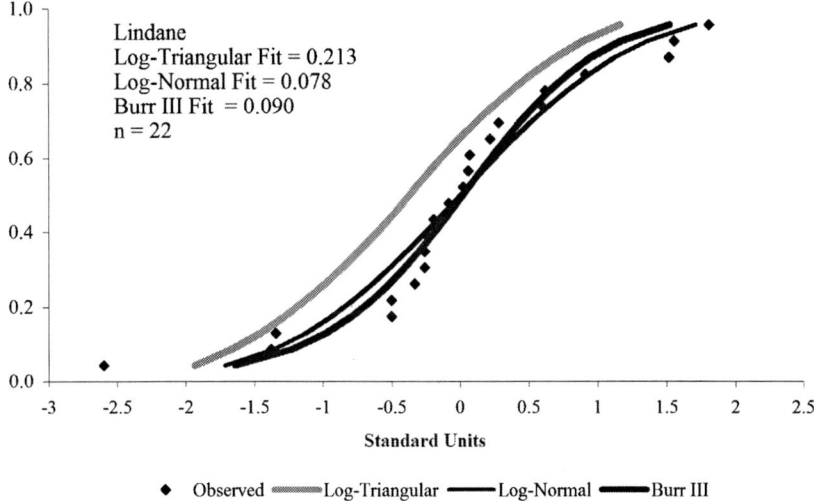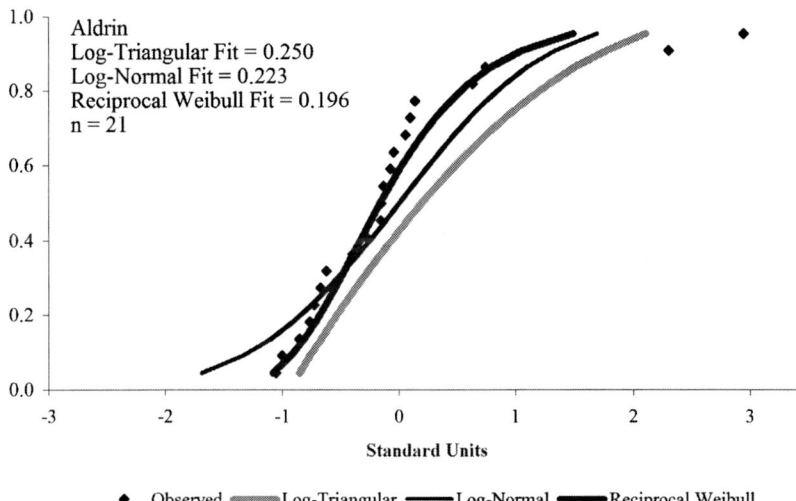

Fig. 2 (Part 3) Comparison of fits of pesticide toxicity data to log-triangular, log-normal, and Burr Type III distributions (including Reciprocal Weibull) for lindane and aldrin

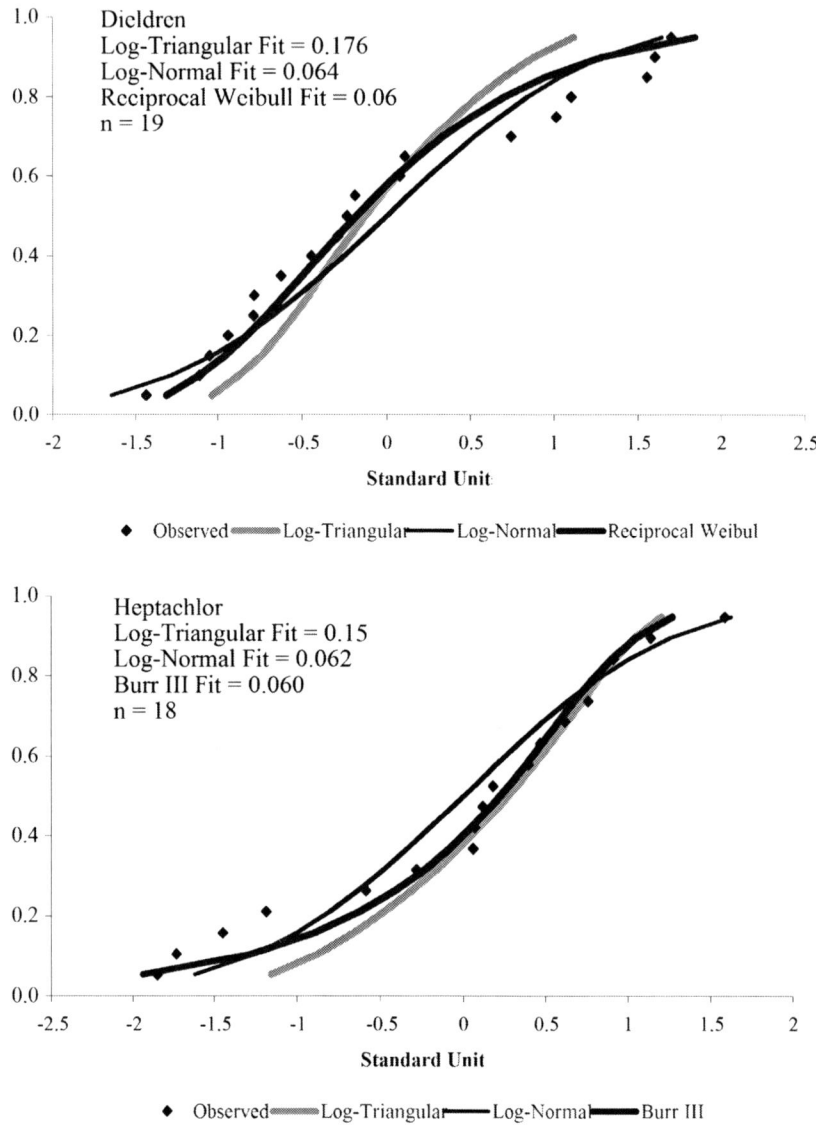

Fig. 2 (Part 4) Comparison of fits of pesticide toxicity data to log-triangular, log-normal, and Burr Type III distributions (including Reciprocal Weibull) for dieldrin and heptachlor

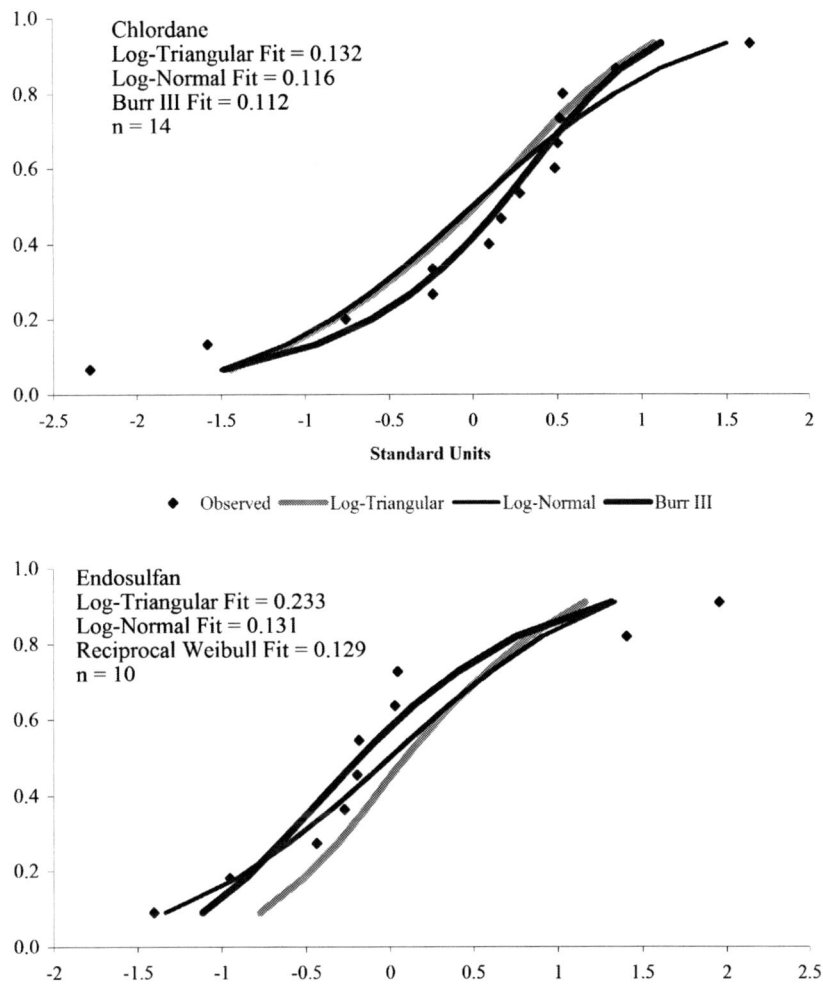

Fig. 2 (Part 5) Comparison of fits of pesticide toxicity data to log-triangular, log-normal, and Burr Type III distributions (including Reciprocal Weibull) for chlordane and endosulfan

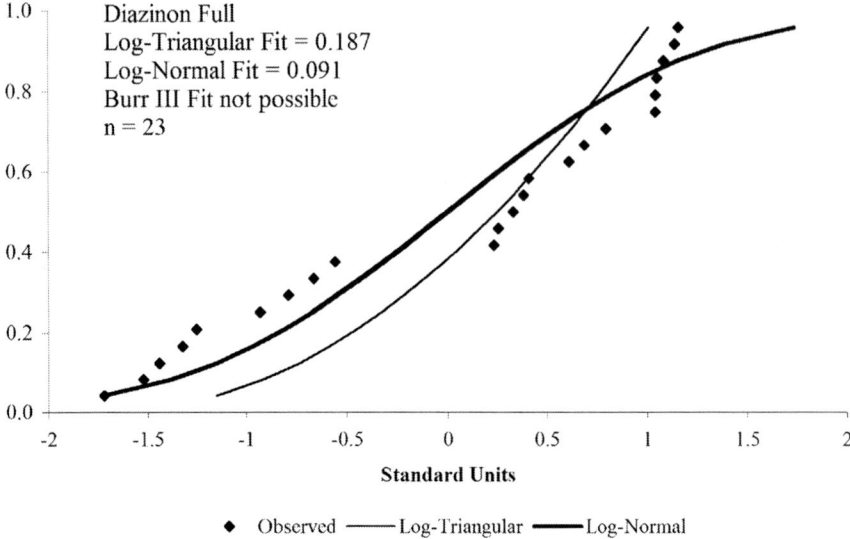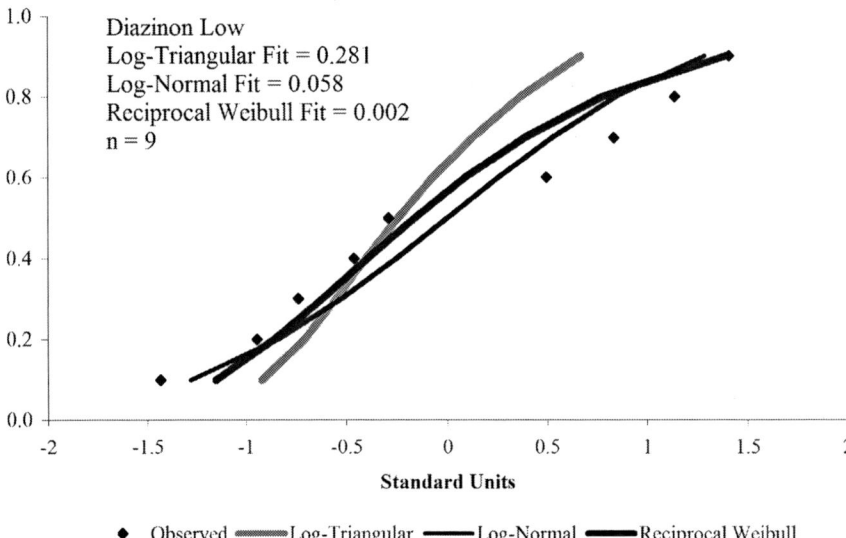

Fig. 2 (Part 6) Comparison of fits of pesticide toxicity data to log-triangular, log-normal, and Burr Type III distributions (including Reciprocal Weibull) for full diazinon and lower subset of diazinon data sets

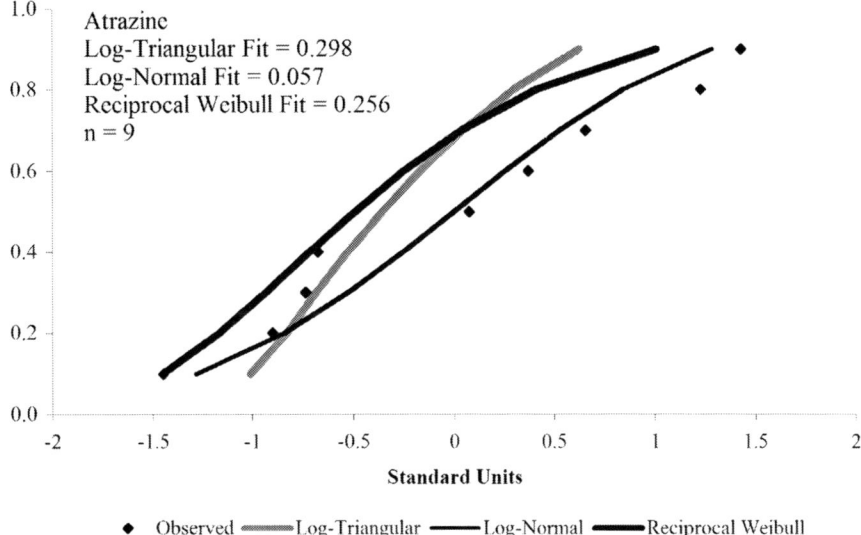

Fig. 2 (Part 7) Comparison of fits of pesticide toxicity data to log-triangular, log-normal, and Burr Type III distributions (including Reciprocal Weibull) for atrazine

As $k \to \infty$, the Burr III distribution approaches the reciprocal Weibull distribution, and as $c \to \infty$, the Burr III distribution approaches the reciprocal Pareto distribution. Thus, these so-called limiting distributions are used in cases where $k > 100$ and $c > 80$, respectively (Campbell et al. 2000; CSIRO 2001).

The cumulative distribution function for the reciprocal Weibull distribution is

$$F(x) = \exp^{-\alpha x^{-\beta}} \qquad (5)$$

where

$F(x)$ = probability of x;
α and β are fit parameters;
x = data point in the distribution.

For the reciprocal Pareto distribution the cumulative distribution function is

$$F(x) = \left(\frac{x}{x_0}\right)^{\theta} \qquad (6)$$

where

$F(x)$ = probability of x;
x_0 and θ are fit parameters;
x = data point in the distribution.

Table 15 Burr III family distribution fit parameters for data sets in Table 14. Parameters c, b, and k apply to the Burr III distributions; α and β apply to the reciprocal Weibull distribution

Pesticide	Distribution	c or α	b or β	k
Chlorpyrifos	Reciprocal Weibull	0.6979	0.3855	--------
DDT	Burr III	0.8576	0.7528	5.077
Toxaphene	Burr III	1.0852	3.6524	2.366
Endrin	Reciprocal Weibull	0.8914	0.7889	--------
Lindane	Burr III	1.552	76.521	0.8204
Aldrin	Reciprocal Weibull	15.908	1.095	-------
Dieldrin	Reciprocal Weibull	6.653	0.7061	-------
Heptachlor	Burr III	2.147	92.256	0.2830
Chlordane	Burr III	3.275	67.499	0.3572
Endosulfan	Reciprocal Weibull	1.743	0.667	-------
Diazinon full	Could not fit	---------	---------	-------
Diazinon low	Reciprocal Weibull	1.454	0.8207	-------
Atrazine	Reciprocal Weibull	34375	1.157	-------

Table 15 shows the results of curve-fitting for the 12 data sets in Table 14. The BurrliOZ v. 1.0.13 program (CSIRO 2001) was used to fit Burr Type III distributions to each data set, except the full diazinon set, which is bimodal (histogram, Fig. 1, Part 4; observed data, Fig. 2, Part 6). Using only the lowest nine diazinon toxicity values, a reciprocal Weibull distribution (one of the Burr III family) was fit.

Erickson and Stephan (1988) utilized the following formula to measure goodness of fit to a number of different distributions:

$$\frac{\sum_{n}(X_R - E(X_R))^2}{\sum_{n}(X_R - \overline{X}_R)^2} \quad (7)$$

where

n = number of data in set;
X_R = observed value of X at rank R;
$E(X_R)$ = expected value of X at rank R;
$\overline{X}_R$ = mean value of X for all ranks.

and

$$E(X_R) = \hat{L} + \hat{S} \bullet E(Z_R) \quad (8)$$

where

$\hat{L}$ = location parameter estimate (mean);
$\hat{S}$ = scale parameter estimate (standard deviation).

Using Equations (7) and (8), the data in Table 14 were tested for goodness of fit to the log-triangular, log-normal, and Burr Type III distributions used in USEPA (1985, 2003a), RIVM (2001), and ANZECC and ARMCANZ (2000), respectively. Results are shown in Table 16 and are included in Fig. 2 (Parts 1–7).

Table 16 Comparison of fit of log-triangular, log-normal, and Burr Type III distributions for data sets from Table 14. Lower number (shaded) indicates better fit. Where both are shaded, fits are equally good. Fit is measured as described in Erickson and Stephan (1988); data were log-transformed prior to analysis

	Log-triangular	Log-normal	Burr Type III
Chlorpyrifos	0.120	0.087	0.049
DDT	0.035	0.033	0.021
Toxaphene	0.153	0.261	0.017
Endrin	0.154	0.149[a]	0.075
Lindane	0.213	0.063	0.047
Aldrin	0.147	0.039	0.057
Dieldrin	0.176	0.069	0.068
Heptachlor	0.151	0.058	0.062
Chlordane	0.132	0.096	0.041
Endosulfan	0.233	0.110	0.069
Diazinon full	0.187	0.107[a]	No fit
Diazinon low	0.281	0.042	0.074
Atrazine	0.298	0.040	0.069

[a]Normal fit rejected by analyses done with the programs ETX (Van Vlaardingen et al. 2004) and JMP (2004).

In any case, the log-triangular distribution is not the best fit, although it does well with the large DDT data set. Eleven of the data sets may be described by log-normal distributions ($p > 0.05$; Fig. 1, Parts 1–4), whereas that of endrin and the full, bimodal diazinon set may not (Fig. 1, Parts 2 and 4, respectively). Burr III distributions were fit to 12 of the data sets, but could not be fit to the full diazinon set. For toxaphene, lindane, dieldrin, and heptachlor, the log-normal and Burr III fits were equally good (ratio of fit numbers <1.5). For six of the sets, the Burr Type III distribution fits better than either the log-triangular or log-normal (Table 16), while the log-normal distribution is the best fit in two cases. When the log-normal fit is better, it is only slightly better (ratio of fit numbers = 1.5–1.8). On the other hand, there are cases where the Burr III fit is much better (ratio of fit numbers = 1.6–15). Although the goodness-of-fit test used in Table 16 indicates that the log-normal distribution is better than the log-triangular for the full diazinon set, Fig. 1 (Part 4) indicates that this set cannot be described by a normal distribution ($p < 0.05$). Thus, the Burr Type III result of "no fit" (Table 16) is more accurate. For the lower portion of the diazinon set, the Burr Type III distribution is the best fit. For endrin, the only distribution that fits is the Burr Type III.

The Burr family of distributions provides a better fit than the log-triangular distribution in all cases tested, and it provides an equivalent or far better fit than the log-normal distribution in most cases. This is expected because the Burr III family of distributions approximates the log-normal and log-triangular distributions (CSIRO 2001). Based on the fit of the SSD alone, the Burr III family of distributions is the best candidate for use in the UCDM for derivation of criteria by the SSD technique. However, there are a few other factors that are important in choosing an SSD technique. SSD methodologies of the USEPA (1985, 2003a), RIVM (2001), and ANZECC and ARMCANZ (2000) are discussed and compared further in Section 3.1.5, and an SSD will be chosen in Section 3.1.6.

3.1.2 Percentile Cutoff

To use an SSD procedure for criteria derivation requires selection of a percentile of the distribution as a cutoff point. This is often interpreted to mean that species lying above this point in the distribution will be protected as long as the concentration of chemical is below the concentration at the selected percentile, but species lying below the percentile would be harmed. Van Straalen and Van Leeuwen (2002) note that it is not correct to interpret the 5th percentile to mean that 5% of species will be harmed (as was argued, for example, by Lillebo et al. (1988), regarding the USEPA 1985 methodology). Rather, this approach is one method for derivation of a predicted no-effect concentration, and although the choice of the 5th percentile is purely a pragmatic one, it has been validated by field studies. Solomon et al. (2001) note that any percentile may be chosen as long as it can be validated against knowledge and understanding of ecosystem structure and function.

The USEPA rationale for choosing the 5th percentile is simply that criteria values, derived using the 10th or 1st percentiles, seemed to be too high and too low, respectively; the 5th percentile was selected because it falls between the 10th or 1st percentiles (Stephan 1985). Using the USEPA methodology (USEPA 1985, 2003a) the chronic criteria are derived directly from the 5th percentile of MATC values. Acute criteria are derived from EC_{50} or LC_{50} data. Since 50% effect is not acceptable, the 5th percentile values are divided by a safety factor of 2 to arrive at the final acute criterion value. This figure was based on 219 acute toxicity tests with various chemicals, which showed that the mean concentration that did not cause mortality greater than control was 0.44 times the LC_{50}. The inverse of 0.44 (2.27) was rounded to 2 for use in USEPA methods. Subsequent studies have shown good agreement between USEPA criteria and no-effect concentrations determined in experimental stream studies (USEPA 1991). The Dutch guidelines (RIVM 2001) use the 5th percentile for derivation of environmental limits. Specific reasons for this choice are not given, but the 5th percentile has been validated against field NOEC values in studies by Emans et al. (1993) and Okkerman et al. (1993). The Australia/New Zealand guidelines (ANZECC and ARMCANZ 2000) consider the question more rigorously, but still arrive at the 5th percentile level for the simple reasons that it works well in the Dutch guidelines (RIVM 2001) and it gives criteria that agree with NOEC values from multispecies tests. The reason for not regularly using a lower percentile is that the uncertainty is very high in the extreme tail of the distribution and the uncertainty can contribute more to derived criteria than the data. However, the Australia/New Zealand guidelines do use the 1st percentile as a default value for high conservation ecosystems, for bioaccumulative substances, and for cases where an important species is not protected at the 5th percentile level. To provide further information to water quality managers in Australia/New Zealand, other percentile levels are also calculated so that criteria are given based on the 1st, 5th, 10th, and 20th percentiles.

Other researchers have also found good correlation between criteria derived from the 5th percentile of single-species SSDs and NOEC values determined in multi-species tests (Hose and Van Den Brink 2004; Maltby et al. 2005; Versteeg et al. 1999). On the other hand, Zischke et al. (1985) found that a laboratory-derived criterion concentration of pentachlorophenol was not protective of invertebrates and fish in outdoor experimental channels. Maltby et al. (2005) determined that concentrations of pesticides derived from the 5th percentile of SSDs with 95% confidence was protective of species in freshwater ecosystems, but concentrations derived with 50% confidence were not protective and required application of a safety factor.

The 5th percentile SSD cutoff, which has been validated against multispecies NOEC values in many cases, is a level that balances the desire to select a percentile near zero with the need to avoid utilizing the highly uncertain tails of the distributions. The UCDM will use the 5th percentile, but since a few studies have shown this level to be underprotective, criteria derived from this value will be evaluated against available data from tests with multispecies, ecosystem, sensitive species, and threatened or endangered species. If evidence suggests that the 5th percentile will not be protective, criteria may be adjusted downward. The recommended means of making such an adjustment is to use either a lower 95% confidence limit estimate of the 5th percentile (see Section 3.1.3), or a median or 95% confidence limit estimate of the 1st percentile.

3.1.3 Level of Confidence

With SSD procedures, it is necessary to decide what level of certainty is desired in the resulting concentration. The USEPA approach (USEPA 1985, 2003a) does not provide a means to determine levels of confidence for derived criteria; all are derived as the median estimate of the 5th percentile, meaning that the true value may be greater or less than the estimated value with equal probability. All other SSD methodologies result in a criterion derived from a specified percentile level and a specified level of confidence. Uncertainty in an extrapolated value results from the risk that the extrapolated value is wrong (Aldenberg and Slob 1993). The distribution around the extrapolated value can be used to determine lower boundaries for the extrapolated value (Aldenberg and Slob 1993; Kooijman 1987; Van Straalen and Denneman 1989; Wagner and Løkke 1991). By evaluating this uncertainty, it is possible to state that the true 5th percentile value falls above (or below) the extrapolated value with, for example, a 50, 90, 95%, or another level of certainty.

These distributional confidence limits assume that the uncertainty in the fitted distribution is the greatest source of uncertainty in the criteria calculation and it will out-weigh other sources of uncertainty, such as

(a) the uncertainty of an LC_{50} or MATC calculation, which could be expressed as the confidence limits of the reported LC_{50};
(b) the uncertainty of the reported toxicity test concentrations from the method in which they were determined;
(c) the effect of the conditions of a particular lab set up or batch of organisms, which is indicated by different studies for the same species that report different LC_{50} values.

The differences in species sensitivities usually vary by several orders of magnitude (as seen in Table 14) and are likely to overshadow the other calculable sources of uncertainty listed above. Because there is large species-to-species variation within the distribution, the uncertainty in the fit of the distribution is likely to be the best available quantitative measure of uncertainty in the criteria.

Although different confidence levels may be calculated, the most statistically robust is the 50%, or median, estimate (ANZECC and ARMCANZ 2000; EVS 1999; Fox 1999). If there is evidence that a median 5th percentile estimate will not be adequately protective, then a lower 95% confidence limit estimate of the 5th percentile may be used instead. Variability in the tails of the distributions tends to compound, rather than clarify, uncertainties. Ultimately, the selection of certainty levels is a policy decision, but one that can be informed by understanding the limitations of values derived from distributional tails. In the UCDM we present a method for derivation of criteria with multiple levels of certainty so that environmental managers can choose values that best suit their needs.

3.1.4 Aggregation of Taxa and Outliers

One challenge in the use of SSDs is to fit the data to an appropriate distribution prior to extrapolation. One way to achieve a better fit is to break data into groups rather than to pool it together in one SSD. Data may be grouped according to toxicant mode of action, habitat (e.g., lentic vs. lotic), reproductive strategy, or life cycle (Solomon and Takacs 2002). TenBrook et al. (2009) concluded that it is important to include all species in the criteria derivation procedure. However, it is reasonable, especially in construction of SSDs, to separate species into groups if multi-modal distributions are evident. Alternatively, if there is no justifiable difference between apparent groups (e.g., vertebrates and invertebrates, or plants and animals) then the data should be pooled for criteria derivation. Because these data are not expected to be normally distributed, no statistical test for outliers will be included. Most of the distributions discussed in this report are adept at handling outliers, so critical examination of the data is the main course of action suggested. Apparent "outliers" should be scrutinized for possible sources of error (typographical errors, units reported incorrectly, inappropriate methods). Inability to fit the distribution is more likely to be caused by a bimodal group of data in which there is not enough data to actually characterize the two subsets. In this case, non-distributional methods (AFs) should be used, which would be similar to the method for limited data sets that do not meet the taxa requirements for using a distributional method. If the distribution is unimodal and cannot be fit with a larger data set, it may be reasonable to exclude outliers, provided that there is some rationale for the difference in sensitivity and that the criteria be adjusted if not protective of sensitive species.

3.1.5 Comparison of Methods

In the previous sections, important specific aspects of SSD procedures have been presented, including the fit of different SSDs, the percentile cutoff, confidence level, and aggregation of data. In this section, the methods of three agencies are more broadly reviewed and compared by deriving example criteria. Two of these methods are the widely used SSD techniques of RIVM (2001) and the USEPA (1985, 2003a). Also included is the ANZECC and ARMCANZ (2000) technique, which represents an improvement over RIVM (2001) in that it provides a way to fit distributions to data sets (a technique supported by OECD 1995a). Results of running all three methods with all data sets (Table 14) are given in Table 17. Before that, each method is described, assumptions are stated, and advantages and disadvantages are listed.

Table 17 Results of analyzing pesticide data sets (Table 14) with SSD methods from USEPA (1985), the Netherlands (RIVM 2001), and Australia/New Zealand (ANZECC and ARMCANZ 2000). All values represent 5th percentiles (not criteria) expressed in μg/L

Pesticide	USEPA (median)	RIVM (median)	RIVM (95th)	A/NZ (median)	A/NZ (95th)	Lowest value in data set[a]
Chlorpyrifos	0.033	0.005	0.0003	0.023	0.018	0.035
DDT	0.845	0.659	0.322	0.97	0.61	0.36
Toxaphene	1.21	1.15	0.590	1.54	1.04	0.8
Endrin	0.20	0.081	0.027	0.22	0.15	0.15
Lindane	2.65	6.82	2.77	7.4	2.18	10
Aldrin	3.76	3.45	1.54	4.59	3.52	4
Dieldrin	2.67	1.62	0.415	3.10	2.22	2.5
Heptachlor	0.768	1.22	0.322	0.67	0.07	0.9
Chlordane	2.10	5.65	2.11	5.21	1.64	3
Endosulfan	0.183	0.188	0.017	0.44	0.24	0.34
Diazinon all	0.449	0.460	0.043	NC	NC	0.3773
Diazinon low	0.260	0.251	0.036	0.41	0.25	0.3773
Atrazine	2,514	2,276	572	3,233	2,477	3,000

[a] From Table 14

A. Assumptions Common to the Three Methods

All three methods use an SSD to extrapolate to the 5th percentile. Several assumptions apply to such SSD procedures and are given here:

(1) surrogate species are good representatives of species of concern;
(2) protecting species from direct adverse effects will also protect them from indirect adverse effects;
(3) effects that occur on a species in laboratory tests will generally occur on the same species in comparable field situations;
(4) extrapolation of the 5th percentile of single-species toxicity values will produce a value that is protective of all species in an ecosystem;
(5) protecting the most sensitive species will protect all species in an ecosystem;
(6) surrogate species represent a random sampling of all species in an ecosystem.

B. USEPA (1985, 2003a)

USEPA methods use a log-triangular distribution to calculate a final acute value (FAV), which is also the 5th percentile estimate. To calculate the FAV, the total number of SMAVs, and usually the lowest four SMAVs, are used. This calculation can be done by hand or using a spreadsheet with equations that are included in TenBrook et al. (2009, see Erickson and Stephan 1988 for derivation). Note that the USEPA (1985, 2003a) SSD procedure is defined in terms of genus mean acute values (GMAVs; see Section 2.7), but SMAVs are used in the UCDM. Thus, for comparison, SMAVs are substituted for GMAVs for the criteria derived in Table 17.

The final chronic value (FCV) may be derived in the same manner if enough chronic data are available; however, the FCV is typically derived by application of an ACR (Section 3.2.5) to the FAV.

Assumptions specific to this method:

(1) No species succumbs to infinitesimal, or tolerates infinite concentrations of toxicant (Erickson and Stephan 1988).
(2) Data sets represent independent random samples from symmetrical log-triangular distributions.
(3) Aquatic ecosystems can tolerate some stress and occasional adverse effects, therefore protection of all species at all times and places is not necessary.
(4) Censored data (data expressed as < or > a value) can be used if not in the lowest four values.

Minimum data values required: 8
Advantages:

(1) Fitting to only the four toxicity values nearest the 5th percentile eliminates problems that arise when toxicity data sets do not meet the log-triangular distribution assumption.

(2) Focuses on sensitive end.

Disadvantages:

(1) There is no biological basis for selecting a triangular distribution (ANZECC and ARMCANZ 2000).
(2) Not all of the data are used to fit the distribution.
(3) No associated confidence levels can be calculated.
(4) Requires the most data of the three methods.

C. RIVM (2001; Formerly MHSPE 1994)

Environmental risk limits (5th percentile estimates) are derived using the SSD procedure of Aldenberg and Jaworska (2000). That is, HC_p values (hazardous concentrations affecting $p\%$ of species) are calculated based on a log-normal SSD. Equations are included in TenBrook et al. (2009) and a computer program called ETX 2.0 is available for making these calculations (Van Vlaardingen et al. 2004; available for free at http://www.rivm.nl/rvs/overig/risico/methoden/ETX.jsp). Calculations are usually done with NOEC data to calculate chronic criteria, but can also be done with acute data.

Assumptions specific to this model:

(1) No toxicity thresholds exist.
(2) Data sets represent independent random samples from symmetrical log-normal distributions.

Minimum data values required: 4
Advantages:

(1) All of the data are used to fit the distribution.
(2) Associated confidence levels can be calculated.
(3) Requires the least data of the three methods.

Disadvantages:

(1) Data sets that do not fit the symmetrical log-normal distributions would be problematic.

D. ANZECC and ARMCANZ (2000)

The Australian/New Zealand guidelines use the same method as do the Dutch, but with a curve-fitting procedure that overcomes the problem of data that do not fit an assumed distribution. Using the program BurrliOZ v. 1.0.13 (Campbell et al. 2000; CSIRO 2001), which is available for free at http://www.cmis.csiro.au/Envir/burrlioz/, data are fit to either the Burr III or one of the limiting distributions, and the median 5th percentile value is calculated. The equations are shown in Section 9.3.2A (Burr 1942, also discussed in Section 3.1.1). As in the Dutch procedure, NOEC values are normally used to calculate a chronic criterion, but this model can also be used with acute data.

Assumptions specific to this model:

(1) No toxicity thresholds exist.
(2) Data sets represent independent random samples from symmetrical Burr Type III distributions.

Minimum data values required: 5
Advantages:

(1) All of the data are used to fit the distribution.
(2) Associated confidence levels can be calculated.

Disadvantages:

(1) Data sets that do not fit symmetrical Burr Type III distributions would be problematic.

One other advantage of this method over the USEPA (1985, 2003a) and RIVM (2001) methods is that if a Burr III distribution, or one of the limiting distributions, cannot be fit to a data set, then no 5th (or other) percentile value can be calculated. It may still be possible to derive a 5th percentile value if modification of the data set is warranted due to bimodality or the presence of outliers. Using either the USEPA and RIVM methods, an unwitting user can determine a 5th percentile value whether or not the distributional assumptions are met. Such values would be unreliable.

E. Results and Discussion of SSD Model Comparison

Table 17 shows the 5th percentile values derived for each of 12 pesticides using each of the three SSD procedures. The USEPA method results in one value, the median estimate of the 5th percentile, whereas each of the others (ANZECC and ARMCANZ 2000; RIVM 2001) result in a median estimate, as well as a lower 95th percentile estimate. The true 5th percentile value has an equal certainty of falling above or below the median estimates, but has a 95% certainty of falling above the lower 95th percentile estimate. Other levels of certainty may be calculated with the ANZECC and ARMCANZ (2000) method as well, but these calculations become less and less reliable in the extreme tails of the distribution.

Many of the 5th percentile values derived by the various methods are similar. For example, the median 5th percentile values derived for DDT, toxaphene, aldrin, dieldrin, heptachlor, diazinon (low), and atrazine are within a factor of 2 by all three methods. The endrin median 5th percentile value is 2.5–2.75 times lower by the RIVM (2001) method compared to the other two, but that data set violated the assumption of log-normality, thus the RIVM (2001) method is not a good choice for the endrin data. The chlorpyrifos median 5th percentile value determined by the RIVM (2001) method is a factor of 4.6–6.6 lower than those obtained by the other methods. Although the chlorpyrifos data fit a normal distribution, the box plot in Fig. 1 (Part 1, B) reveals that the distribution is right-skewed, which would cause the 5th percentile value to fall at a lower value than if the distribution were not skewed. The 5th percentile values obtained by USEPA (1985, 2003a) and ANZECC and ARMCANZ (2000) for endrin are very similar, in spite of the skew of the data. For lindane, the median 5th percentile value obtained from the USEPA method is less than half that obtained by the other two methods. This is likely due to the presence of a low outlier in the lindane set (see Fig. 2, Part 2; outliers were left in for this analysis because they are part of the final USEPA criteria sets). As discussed in the TenBrook et al. (2009) and GLEC (2003), low outliers tend to lower the 5th percentile values derived by the USEPA method. The same phenomenon occurs for chlordane, in which the USEPA value is less than half that of the other two. For endosulfan, the ANZECC and ARMCANZ (2000) method produces a median value that is more than twice the values obtained by the other two methods. In this case, Fig. 1 (Part 3) indicates a high outlier, which is handled better by the Burr III family of distributions than either the log-triangular or log-normal ones (Table 16).

An important consideration in assessing the various methods is whether or not they will produce protective criteria. The acute criterion will be derived by dividing the 5th percentile value by 2, so it is of interest to compare the lowest value in each data set with the resulting criterion. The last column of Table 17 shows the lowest value from each data set. For most of the pesticides, criteria derived from any of the median 5th percentile values would be lower than the lowest value. However, for DDT the median values derived by the USEPA (1985, 2003a) and ANZECC and ARMCANZ (2000) methods would result in criteria of 0.42 and 0.48 μg/L, respectively, which are higher than the lowest value of 0.36 μg/L. In such a case, the Australia/New Zealand approach provides the option of using the 95th percentile value to produce a number below the lowest value. By the USEPA method, an additional safety factor would have to be applied to derive a protective criterion. The UCDM includes a step of checking derived criteria against available data and adjusting the criteria if they do not appear to be protective (Section 5).

3.1.6 SSDs in the UCDM

The ANZECC and ARMCANZ (2000) methodology, utilizing the Burr Type III distributions with the BurrliOZ program, offers the best combination of best fit, data requirements, appropriate distributional assumptions, and flexibility in choosing protection and confidence levels as discussed in these last several sections. In addition, all of the distributions historically used for SSD analysis are considered by using the Burr III family: the Burr Type III distribution approximates the log-triangular and log-normal distributions, and the log-logistic distribution is a special case of the Burr III distribution (CSIRO 2001).

Although the ANZECC and ARMCANZ (2000) methodology does have advantages, it is noteworthy that all of the methods currently in use appear to derive protective criteria. In The Netherlands, the log-normal distribution was selected over a log-log distribution (Aldenberg and Slob 1993) because, although the distributions are not very different and results obtained are not different, the normal distribution provides powerful statistical tools (RIVM 2001). Moreover, the OECD (1995a) concludes that the log-normal, log-logistic, and log-triangular distribution methods give very similar results. Here too it was found that, with some exceptions, median 5th percentile values obtained by the ANZECC and ARMCANZ (2000) method are comparable to those obtained by the USEPA (1985, 2003a) methodology.

The UCDM uses the ANZECC and ARMCANZ (2000) SSD procedure, which relies on the BurrliOZ program to calculate 5th percentile values with 50 and 95% confidence that the true 5th percentile value lies above the derived value. The acute criterion derived by the SSD procedure is equal to the median 5th percentile value divided by 2. The safety factor of 2 is applied because the SSD is constructed with toxicity values that indicate a 50% effect level (Section 3.1.2). A 5th percentile value derived from chronic data is the chronic criterion without further adjustment.

3.2 AF Procedure

When fewer than five data values from an appropriate assortment of taxa are available, the SSD procedure cannot be used for criteria derivation. In such cases, an AF method must be used. As TenBrook et al. (2009) stated, AFs are recognized as a conservative approach for dealing with uncertainty in assessing risks posed by chemicals (Chapman et al. 1998). AFs (also called safety factors, application factors, extrapolation factors) are usually applied to account for a wide range of possible effects and situations for which no data exist, including the following: lack of tests with relevant species; persistence or bioaccumulative potential of substances; genotoxic potential; laboratory to field extrapolation; acute-to-chronic extrapolation; variations in mesocosm types for multispecies tests; absence of most sensitive species in multispecies tests; mixture effects; experimental variability; and lack of data (TenBrook et al. 2009). Additional factors may occasionally be applied, based on the professional judgment of the risk assessor. In all cases, the more the toxicity data that are available for species of different trophic levels, different taxonomic groups, and different lifestyles, the smaller the applied factor.

In Table 18 we summarize AFs used by the methodologies reviewed by TenBrook et al. (2009). They range from 1 to 1,000 or more and, with the exception of ACRs, are primarily based on the premise that 10 is a widely used safety factor

Table 18 Assessment factors used in existing methodologies

Methodology	Range of factors	Applied to	Reference
Australia/New Zealand	Acute: NA Chronic: 10–1,000 Default ACR: 10+	Acute: NA Chronic: NOEC (single- or multispecies)	ANZECC and ARMCANZ (2000)
California (draft)	Acute: NA Chronic: 10	Acute: NA Chronic: LOEC	Lillebo et al. (1988)
Canada	Acute: NA Chronic: 10–100+ Default ACR: 10	Acute: NA Chronic: LC/EC_{50}, LOEC	CCME (1999)
European Union/ Denmark	Acute: NA Chronic: 1–1,000 Default ACR: 10	Acute: NA Chronic: LC/EC_{50}, NOEC, QSAR estimates	OECD (1995a), Samsoe-Petersen and Pedersen (1995), ECB (2003)
France	Acute: NA Chronic: 1–1,000	Acute: NA Chronic: LC/EC_{50}, NOEC	Lepper (2000)
Germany	Acute: NA Chronic: 10-1000 Default ACR: 10	Acute: NA Chronic: LC/EC_{50}, NOEC	Irmer et al. (1995)
Great Lakes	Acute: 4.3–21.9 Chronic: NA Default ACR: 2 for FACR 18 for SACR	Acute: LC/EC_{50} Chronic: NA	USEPA (2003a)
North Carolina	Acute: 3 Chronic: 1 Default ACR: 100 for $t_{1/2}$ >96 h 20 for $t_{1/2}$ <96 h	Acute: LC_{50} Chronic: MATC	North Carolina DENR (2003)
South Africa	Acute: 1–100 Chronic: 1–1,000	Acute: LC_{50} Chronic: Final Acute Value	Roux et al. (1996)
Spain	Acute: NA Chronic: 1–100+	Acute: NA Chronic: LC/EC_{50}, NOEC	Lepper (2000)
The Netherlands	Acute: NA Chronic: 1–1,000	Acute: NA Chronic: LC/EC_{50}, NOEC	RIVM (2001)
United Kingdom	Acute: 2–10 Chronic: 1–100	Acute: LC/EC_{50} Chronic: NOEC	Zabel and Cole (1999)
USEPA	Acute: 2 Chronic: NA Default ACR: 2	Acute: Final Acute Value Chronic: NA	USEPA (1985)
USEPA	Acute: 100–1,000 Chronic: 10 Default ACR: 10	Acute: LC/EC_{50} Chronic: MATC	Nabholz (1991)

in toxicology and should be applied for each step of an extrapolation from data at hand to real-world application.

3.2.1 Appropriate Use of AFs

An important point in using AFs is that application of empirically based factors to toxicity data does not quantify uncertainty, but does reduce the probability of underestimating risk. Moreover, the use of AFs also greatly increases the possibility of overestimating risk (Chapman et al. 1998). It is worth restating and evaluating some of the specific points by Chapman et al. (1998) regarding the use of AFs, keeping in mind that each of the points needs evaluation in the context of water quality criteria derivation, as opposed to ecological risk assessment. Ecological risks are assessed based on a specific set of exposure and effects data, usually for a specific site. Numeric water quality criteria are derived considering only effects data and a few exposure factors that directly affect toxicity. Criteria must protect aquatic life, and therefore must err on the side of safety when data are lacking. Criteria may be site-specific, but more often must be valid for a range of sites. When data are lacking, criteria will likely represent an overestimation of risk. More data are required if protective extrapolated values are to approach true values. Hence, the cogency of the points raised by Chapman et al. (1998):

(1) Data supersede extrapolation; that is, if data are available, they should be used.
    This point reinforces the idea that more data will result in better estimates of risk, and therefore, better estimates of appropriately protective criteria.
(2) Extrapolation requires context; use of AFs should be based on existing scientific knowledge.
    This statement is true, but somewhat contradictory. In fact, AFs are used to fill gaps in scientific knowledge. With the exception of measured ACRs (Section 3.2.5), many existing criteria derivation methodologies use standardized factors of 10, 20 and 100, despite lack of supporting data (Chapman et al. 1998). Further, AFs are often based on policy rather than empirical science. One methodology that has employed existing scientific knowledge is that of the Great Lakes (USEPA 2003a), which uses empirically derived factors and default ACRs. Details of the Great Lakes factors are discussed in Section 3.2.4, and the approach used by the Great Lakes method has been adapted for the UCDM.
(3) Extrapolation is not fact; estimates of effect levels obtained using AFs should only be used as screening values, not as threshold values (criteria).
    All criteria are extrapolated values, and although those obtained by application of large factors to small data sets have a high level of conservatism and uncertainty, it is a policy decision whether or not to use them as threshold values. The UCDM includes an AF procedure to be used for data sets that are inadequate for SSD analysis. Less conservative, more certain numbers may be derived if more data become available.

(4) Extrapolation is uncertain; AFs should encompass a range rather than being a single value.

With the exception of the German methodology (Irmer et al. 1995), all existing methodologies utilize ranges of factors, and in all cases factors get smaller as data sets get larger. The AF procedure in the UCDM includes a range of factors.

(5) All substances are not the same; AFs should be scaled relative to different substances, potential exposures, and nature of effects.

As noted in the response to point 4, ranges of factors are used by nearly all existing methodologies. Among the reasons for using larger factors are lack of data, persistence, bioaccumulative potential, mixture toxicity, and potential for genotoxic effects. In effect, AFs for each of these variables achieves the scaling suggested by Chapman et al. (1998). Another example of how this is done is that of the state of North Carolina, which utilizes different default ACRs depending on the half-life of the chemical of concern (North Carolina DENR 2003; Table 18). The UCDM includes other means of addressing bioaccumulative potential, bioavailability, and mixture toxicity and therefore does not incorporate these elements into AFs. Genotoxic effects are of concern in human health risk assessment, but not for protection of aquatic life (see Section 2.1.3 regarding non-traditional endpoints), and thus are not incorporated into AFs either.

(6) Unnecessary overprotection is not useful; AFs for individual extrapolation steps should not exceed 10, and may be much lower.

This statement regarding overprotection is true, but should not be addressed by limiting the size of AFs. Chapman et al. (1998) found ACRs ranging from 1 to 20,000. Thus, for acute-to-chronic extrapolation, factors should definitely not be limited to 10. Also, empirically derived factors, such as those in the Great Lakes methodology (USEPA 2003a), should not be limited to 10. The best way to minimize overprotection is to expand available acute and chronic data sets.

3.2.2 Toxicity Values

One of the choices to be made in using the AF procedure is what kind of toxicity data to use. For the UCDM, separate acute and chronic criteria are derived. Therefore, acute criteria are derived from LC_{50} or EC_{50} data. The situation is not so simple for chronic criteria. As shown in Table 18, most methodologies from the United States (Nabholz 1991; North Carolina DENR 2003; USEPA 1985, 2003a) use MATC values, while most other methodologies from around the world use NOEC values for derivation of chronic criteria; two utilize LOEC values (CCME 1999; Lillebo et al. 1988). The Netherlands (RIVM 2001) and the EU risk assessment technical guidance document (ECB 2003) accept EC_{10} values to represent NOEC values. As presented in Section 2.1.2, hypothesis test results can be used in criteria derivation, providing results are evaluated with respect to test design, MSD (type II error), and effect levels at the NOEC and LOEC. The

MATC is the hypothesis test value used in the UCDM to derive the chronic criteria (Section 2.1.2).

3.2.3 Magnitude Factors

The magnitude of AFs ranges widely among existing methodologies (Table 18). Aside from measured ACRs, little to no justification is given for the magnitude of most AFs (TenBrook et al. 2009). The exception is the Great Lakes methodology (USEPA 2003a), which utilizes FAV factors and default ACRs that are empirically based and theoretically supported (Host et al. 1995); we use Host et al. (1995) methods to derive factors for the UCDM. These factors are discussed further in the following sections.

3.2.4 Acute Factors

Most AF procedures utilize standard factors that get progressively larger as more extrapolation steps are required. Such factors are widely used despite having little scientific basis (Chapman et al. 1998). The acute toxicity factors used in the Great Lakes methodology are based on the work by Host et al. (1995), in which they described both empirical and theoretical methods for derivation of factors that use data sets for all chemical types and are based on log-triangular data distributions. Host et al. (1995) methods were adapted to derive acute factors for the UCDM. The major differences in the approach of the UCDM compared to that described by Host et al. (1995) are the use of only pesticide data (vs. other contaminant types), and the use of Burr Type III distributions (rather than log-triangular distributions). The Host et al. method requires that full data sets be available for several compounds, thus at this point it is only possible to derive acute (vs. chronic) factors by this method.

The goal of the AF process is to allow users to estimate a 5th percentile value when fewer than five data values are available. The magnitude of the AFs must be set to achieve this. The factors are used as divisors for the lowest value in data sets that contain only 1–4 toxicity values.

The following procedure, based on Host et al. (1995) methodology, was applied to each individual pesticide data set (Table 14):

(1) Ninety-nine (99) subsets of five toxicity values were randomly selected with the restriction that the first value had to be for an invertebrate from the genus *Daphnia, Ceriodaphnia,* or *Simocephalus.* These organisms were required because they are known to usually be among the most sensitive and because data for one these species is required by USEPA for pesticide registration. Each successive sample had to fulfill a different requirement in the SSD minimum data set (Section 2.6). The selection of which family to use for the second and subsequent toxicity values in each subset was made randomly.
(2) For each subset of five toxicity values, subsets of 1–4 toxicity values were also created, resulting in 99 subsets of 1–5 toxicity values.

(3) The lowest acute value in each subset of size 1–5 toxicity values was used as the numerator for calculating the AF.
(4) Each of the 99 five-sample subsets was used to generate 5th percentile values using BurrliOZ v. 1.0.13 (Campbell et al. 2000; CSIRO 2001).
(5) The geometric mean of the 99 5th percentile values was used as the denominator for calculating the AF.
(6) This procedure yielded 99 factors for each subset size.
(7) The 95th percentile of the 99 factors was determined for each subset size.

This procedure was followed for chlorpyrifos, DDT, toxaphene, endrin, lindane, aldrin, dieldrin, heptachlor, chlordane, and endosulfan. Atrazine was not included among this group because its data set is from a draft document. Diazinon was not included because the above procedure could not be applied to the data set available for it. The bimodality of the full diazinon set resulted in many subsets that could not be fit to a Burr Type III distribution, and the lower portion of the diazinon data did not include representatives from all five families required for the minimum SSD data set.

In accordance with USEPA (2003a), 95th percentile factors (from step 7) for all pesticides were compiled and the median of those factors for each subset size was selected to be the summary AF (i.e., a factor to apply to all pesticides; Table 19). The summary factors for each sample size are shown in Table 20, along with the estimated 5th percentile toxicity values obtained for each pesticide using the summary factors together with the geometric mean lowest value for each subsample size. Diazinon was not used to derive factors. However, the diazinon data set was used to estimate 5th percentile toxicity values using the factors. In all cases, the five-sample factor produced an estimated 5th percentile value that is comparable to, or below the median 5th percentile value, determined from applying the SSD procedure to the full data set. The four-, three-, and two-sample factors all produced estimated 5th percentile values below the value generated by the SSD procedure. Most of the

Table 19 Compilation of 95th percentile of factors for subsets of 1–5 samples. The median values in the last row are the summary factors for each sample size

Subset size	5	4	3	2	1
Chlorpyrifos	5.56	11.81	52.95	52.95	52.95
DDT	4.08	4.08	4.08	4.08	4.08
Toxaphene	3.86	5.93	8.01	8.01	8.01
Endrin	3.65	5.1	8.49	148.49	1633.43
Lindane	4.03	4.12	4.37	18.83	61.48
Aldrin	5.13	5.13	5.32	5.32	5.32
Dieldrin	3.24	3.24	7.64	63.86	74.95
Heptachlor	3.48	16.3	32.71	97.02	97.02
Chlordane	2.23	3.86	5.94	8.62	8.62
Endosulfan	2.02	13.66	22.04	522.68	1550.24
Median	3.8	5.1	7.8	36	57

Table 20 Median 5th percentile toxicity value estimates for sample sizes of 1–5 acute toxicity values using summary pesticide assessment factors. All values are in μg/L

Sample size	5	4	3	2	1	1	SSD 5th percentile (median)	Lowest value in data set[a]
Factor	3.8	5.1	7.8	36	57	57 × 10		
Chlorpyrifos	0.01	0.012	0.009	0.003	0.002	0.0002	0.022	0.035
DDT	0.25	0.20	0.13	0.03	0.02	0.002	0.97	0.36
Toxaphene	0.68	0.61	0.56	0.18	0.22	0.022	1.54	0.8
Endrin	0.10	0.09	0.08	0.03	1.80	0.18	0.22	0.15
Lindane	4.0	3.4	2.9	1.1	9.2	0.92	7.4	10
Aldrin	1.8	1.5	1.3	0.40	0.48	0.048	4.59	4
Dieldrin	1.2	1.0	0.82	0.40	3.5	0.35	3.10	2.5
Heptachlor	0.36	0.36	0.44	0.24	1.1	0.11	0.67	0.9
Chlordane	2.2	2.1	2.0	0.63	1.0	0.10	5.21	3
Endosulfan	0.09	0.09	0.09	0.06	4.6	0.46	0.44	0.34
Diazinon	0.22	0.22	0.22	0.22	0.22	0.022	0.20	0.3373

[a]From Table 14.

one-sample factors produced similar results. The exceptions are endrin and endosulfan, for which the one-sample factor overestimated the 5th percentile value by eight to tenfold. This occurred because, for those two pesticides, the family Daphniidae was the most tolerant and the one-sample subsets had very high toxicity values. To ensure that criteria derived by the AF procedure are protective, even when based on a single toxicity value, an additional factor of 10 should be assessed. As shown in Table 19, this brings the endrin and endosulfan one-sample values very near the SSD median value. The additional factor of 10 will lead to very conservative criteria in cases where Daphnids are among the most sensitive species, but such conservatism is reasonable when relying on a single datum to make predictions for an ecosystem.

To derive an acute criterion by this method, the lowest value from an acceptable data set is divided by the appropriate factor from Table 19 (depending on sample size). The resulting value represents an estimate of the median 5th percentile value and is divided by 2 to determine the acute criterion. In all cases, for all pesticides shown in Table 19, criteria derived using the proposed factors are below the lowest values in each data set and would be expected to be protective.

3.2.5 Acute-to-Chronic Ratios (ACRs)

If at least five chronic data values are available from five different families, the SSD procedure should be used to derive chronic criteria. However, when chronic data are lacking, the use of ACRs is necessary to extrapolate from acute to chronic toxicity. The ACR is calculated by dividing an acute LC/EC$_{50}$ value by a chronic value (e.g., MATC) derived from the same test, or from tests conducted by the same laboratory under identical conditions (USEPA 1985, 2003a). There are three basic approaches to deriving ACRs: (1) derive chemical-specific, multispecies ACRs using acute and chronic values derived from the same tests (ANZECC and ARMCANZ 2000; USEPA 1985, 2003a); (2) derive chemical-specific, multispecies ACRs using whatever chronic data are available, combined with one or more default ACR values (USEPA 2003a); and (3) use a default ACR value for all chemicals. As these approaches represent a stepwise procedure depending on available data, all are appropriate for inclusion in the UCDM.

A. Single-Chemical, Multispecies ACR Based on Measured Data

The first approach is used in both the USEPA methodologies (USEPA 1985, 2003a) and in the Australia/New Zealand methodology (ANZECC and ARMCANZ 2000). However, only the USEPA methodologies give clear guidance for the procedure. The Great Lakes guidance, which is updated from the 1985 version, is presented here. The procedure requires acute and chronic data from organisms in at least three different families including a fish, an invertebrate, and at least one other acutely sensitive species. If there are not enough freshwater data to fulfill the ACR data requirements, then saltwater species may be used because freshwater and saltwater ACRs have been shown to be comparable (USEPA 1985), and this approach has been accepted in numerous criteria derivations (Siepmann and Finlayson 2000; USEPA 1980a, b, c, d, 2003d, 2005a). For each chronic value (MATC) having at least one corresponding appropriate acute value, an ACR is calculated by dividing the geometric mean of all acceptable flow-through acute tests by the chronic value. Static tests are acceptable for midges, daphnids and other zooplankton. For fish, the acute test(s) should be conducted with juvenile or younger fish. For all species, the acute test(s) should be part of the same study and use the same dilution water as the chronic test. If acute tests were not conducted as part of the same study, but were conducted as part of a different study in the same laboratory and dilution water, then they may be used. If no such acute tests are available, results of acute tests conducted in the same dilution water in a different laboratory may be used. If no such acute tests are available, an ACR is not calculated by this method.

The species mean ACR (SMACR) is calculated for each species as the geometric mean of all ACRs available for that species. For some materials, the ACR seems to be the same for all species, but for other materials the ratio seems to increase or decrease as the SMAV increases. Thus, the multispecies ACR can be obtained in one of three ways, depending on the data available:

(1) If the SMACR seems to increase or decrease as the SMAVs increase, the ACR is calculated as the geometric mean of the ACRs for species whose SMAVs are close to the acute 5th percentile value.

The USEPA methodologies (USEPA 1985, 2003a) do not define what is meant by "SMAVs close to" the 5th percentile value. A definition for use in the UCDM can be developed based on the second approach to derivation of interspecies ACRs (item 2 below), which uses the geometric mean of SMACRs, providing they are within a factor of 10 of each other. Thus, it is reasonable to define species with "SMAVs close to" the 5th percentile as those whose SMACRs are

within a factor of 10 of the SMACR of the species whose SMAV is nearest the 5th percentile value.
(2) If no major trend is apparent and the ACRs for all species are within a factor of 10, the ACR is calculated as the geometric mean of all of the SMACRs.
(3) If the most appropriate SMACRs are less than 2.0, and especially if they are less than 1.0, acclimation has probably occurred during the chronic test. In this situation, the final ACR should be assumed to be 2.0, so that the chronic criterion is equal to the acute criterion.

If the available SMACRs do not fit one of these cases, use the procedure described in Section 3.2.5B to derive an ACR based partially on measured values and partially on assumed values.

B. Single-Chemical, Multispecies ACR Based on Measured and Assumed Values

If insufficient data are available for calculation of an ACR by the preceding procedure, then an ACR is derived by using any available measured ACRs plus enough default ACRs to give a total of three ACRs (USEPA 2003a). For example, if no measured ACRs are available, then three default ACRs are used, of which the geometric mean would be the default ACR itself. If two measured values are available, then just one default ACR is added to the data set. The magnitude of the default ACR is discussed in the following section.

C. Default ACRs

TenBrook et al. (2009) concluded that there is no evidence that default ACR values are appropriate for pesticides, in general. Nonetheless, adequate chronic data are often not available and some means of estimating an ACR is needed. The Great Lakes guidance uses a default ACR of 18 (USEPA 2003a), which represents the 80th percentile value of all available ACRs from USEPA criteria documents (Host et al. 1995). This seems a reasonable approach because it is based on ACRs that have been derived from carefully reviewed studies. Some of the very high ACRs reported in the literature have been rejected by USEPA upon such review (e.g., diazinon ACRs determined by Kenaga 1982). For the UCDM, the default ACR used in the Great Lakes guidance was recalculated to include only pesticide data from Host et al. (1995), as well as the ACR in the CDFG diazinon criteria document (Siepmann and Finlayson 2000), and a new chlorpyrifos value calculated by the UCDM. The results of this calculation are presented in Table 21. Based on this data set, a default ACR of 12.4 was calculated for use in the UCDM, with the following caveats: (1) if data sets collected according to the UCDM lead to different ACR values, those values may be substituted into this table and the default ACR recalculated; (2) if previously calculated ACRs are shown to be invalid based on data sets collected according to the UCDM, then those values should be removed and the default ACR recalculated;

Table 21 Calculation of default ACRs

Chemical	ACR
Chlordane[a]	14
Chlorpyrifos[b]	2.2
Diazinon[c]	3.0
Dieldrin[a]	8.5
Endosulfan[a]	3.9
Endrin[a]	4.0
Lindane[a]	25
Parathion[a]	10
80th percentile	12.4

[a]Taken from Host et al. (1995); originally from USEPA criteria documents.
[b]Derived using the UCDM.
[c]Siepmann and Finlayson (2000).

and (3) if additional pesticide ACR values become available, the default ACR should be recalculated.

The chronic criterion is calculated by dividing the acute 5th percentile value (derived by SSD procedure or estimated by AF procedure) by the ACR (derived by one of the three methods in the Sections 3.2.5A, 3.2.5B, and 3.2.5C):

$$\text{Chronic Criterion} = \text{5th Percentile Value} \div \text{ACR} \qquad (9)$$

3.3 Averaging Periods

Criteria derived according to either the SSD or AF procedures are expressed in terms of how much of a chemical may be in the water without causing harm (i.e., in terms of magnitude), but without consideration of for how long (duration) or how often (frequency) that level may be exceeded without harm. Section 3.4 addresses the frequency component. This section explores the question of duration.

Derivation of separate acute and chronic criteria, as is done in the UCDM, provides a duration component, but criteria derived from studies conducted under constant exposure scenarios do not account for the possibility of pulsed, or otherwise uneven, exposures. Such exposures are common in the Sacramento and San Joaquin River basins (TenBrook et al. 2009) and criteria need to reflect that fact. Time-to-event models could potentially provide a way to express criteria for any given exposure duration, but as discussed in Section 2.1.6, such models are not currently feasible for use in criteria derivation. The best readily usable approach for determination of an appropriate duration component is to consult the literature. This is the approach used by the USEPA (1985, 2003a) to set both acute and chronic averaging periods. The averaging period is the period of time over which the receiving water concentration is averaged for comparison with criteria concentrations (USEPA 1994). There are two aspects to consider in setting an averaging period. First is to set the period long enough such that toxicity might occur due to an exceedance, and second is to set the period short enough that the effects of concentration fluctuations on the average concentration are minimized. For example, the USEPA (1985, 2003a) sets the acute averaging period at 1-h based primarily on the fact that ammonia exerts its effects in that time frame, but also because if the period were longer, peak concentrations would be masked in the averaging process. Similarly, 4–7 days has been shown to be long enough to observe the equivalent of chronic toxicity (USEPA 2002b), but short enough to minimize the effect of concentrations fluctuations (USEPA 1991). The 4-day averaging period used in the USEPA methodologies is reasonable and is used in the UCDM (TenBrook et al. 2009). The 1-h average is based on ammonia toxicity and may not be appropriate for pesticide criteria, a consideration to be explored in the following section.

To establish an appropriate acute averaging period for the UCDM, a literature review was conducted to try to determine the time course of acute pesticide toxicity. Although many studies include reports of pesticide effects after very short toxicant exposures, only a handful consider environmentally relevant concentrations. Researchers often expose organisms to concentrations that are many-fold higher than the 96-h LC_{50} values determined in continuous exposure tests, or at concentrations higher than would ever be expected in the environment (e.g., Barry et al. 1995a, b; Jarvinen et al. 1988; Naddy et al. 2000; Peterson et al. 2001).

Among studies in which environmentally relevant concentrations were considered, Cold and Forbes (2004) reported adverse effects on survival and reproduction in *Gammarus pulex* after 1-h pulses of esfenvalerate as low as 0.05 μg/L, with effects observed for as long as 2 weeks after the exposure. Forbes and Cold (2005) similarly reported effects on larval survival and development rates of the midge *Chironomus riparius* with esfenvalerate exposures as brief as 1 h. Heckman and Friberg (2005) found similar results with pulsed exposures of streams to lambda-cyhalothrin. A 1-h pulse application of lambda-cyhalothrin at ≥0.1 μg/L caused a nearly instantaneous increase in drift of stream macroinvertebrates, with increases occurring within 2 h at concentrations ≥0.01 μg/L (Lauridsen and Friberg 2005). In that same study, *G. pulex* drift increased within 3-h after application at 0.001 μg/L. Holdway et al. (1994) reported significant 96-h mortality to larval rainbowfish (*Melanotaenia fluviatilis*) with 1-h pulse exposures to esfenvalerate at 0.32 μg/L. Significant mortality occurred in 2 days to *D. magna* exposed to a 24-h pulse of fenvalerate at 3.2 μg/L (Reynaldi and Liess 2005). For the organophosphates pirimiphos-methyl and temephos, Brown et al. (2002) reported 24-h LC_{50} values from 1-h pulse exposures of rainbowfish (*Melanotaenia duboulayi*) that were lower than estimated environmental concentrations. Schulz and Liess (2001) reported reductions in emerged individuals of the insect *Limnephilus lunatus* at 154 days, as well as reduction in dry wt and reduction in biomass at 240 days after 1-h pulse exposures to fenvalerate. Sublethal responses were observed at 0.001 μg/L and lethal effects at 0.1 µg/L (Schulz and Liess 2001). In tests of 1-h vs. 10-h equivalent doses of fenvalerate (measured in µg h), Schulz and Liess (2000) reported long-term effects on emergence success and dry weight of caddisfly larvae, with significantly stronger effects from the stronger 1-h pulses vs. the lower level 10-h pulses. Chronic lethal and sublethal effects were observed in *D. magna* after a 24-h pulse exposure of fenvalerate, with complete mortality occurring after 10 days at 3.2 µg/L (Reynaldi and Liess 2005).

Data from intermediate time points in standard toxicity tests is another type of data that can be considered in the analysis of averaging periods. This type of data is regularly collected, but rarely reported. Several studies of acute toxicity of chlorpyrifos and diazinon with *Neomysis mercedis* (mysid), *Ceriodaphnia dubia*, and *Physa* spp. (pond snail) by the CDFG (CDFG 1992a to k, 1998a, b), and one of diazinon with *Lepomis macrochirus* (bluegill) by CIBA-GEIGY (1987) were obtained for review. Results from all of these studies indicate that chlorpyrifos and diazinon are not particularly fast-acting toxicants, and mortality occurred at each 24-h observation period throughout the test (as opposed to only at the earliest observation period).

Based on the available pesticide literature, the 1-h acute averaging period utilized by the USEPA (1985, 2003a) is reasonable for the UCDM. This value is conservative for some pesticides that do not exert effects quickly (e.g., chlorpyrifos and diazinon), but should be protective for pyrethroids that exhibit latent effects after very short acute exposures, and for organophosphates (e.g., pirimiphos-methyl and temephos) that have been shown to exert their effects with 1-h pulse exposures of environmentally relevant magnitude.

To summarize, criteria derived by the UCDM include an expression of allowable exposure duration. For acute criteria, a 1-h averaging period is established, while for chronic criteria a 4-day period is established.

3.4 Allowable Frequency of Exceedance

In addition to magnitude and duration, it is necessary to consider the frequency with which a pesticide concentration may exceed a criterion without causing harm to aquatic organisms. For the UCDM, the allowable frequency of exceedance is based on a review of studies of the ability of organisms to recover from brief exposures to pesticides. Generally, studies of post-exposure recovery are addressed to one of several levels of organization: ecosystem, community, population, species, and individual. All levels should be considered to maintain waters free of toxic substances in concentrations that produce detrimental physiological responses in aquatic life.

3.4.1 Review of the Literature

Yount and Niemi (1990) provide a good review of studies of recovery of lotic communities and ecosystems following physical or chemical disturbances or stresses. They make an important distinction between press disturbances that occur over long periods of time and cause sustained alterations in ecosystems vs. pulse disturbances that occur over shorter periods and cause brief ecosystem alteration. The allowable duration of criteria exceedances included in the UCDM (1 h for acute criteria; 1 h to 4 days for chronic criteria) are pulse disturbances by this definition. Thus, the same allowable frequency of exceedance applies to both acute and chronic criteria. This review focuses on studies of pulse, rather than press, disturbances. To examine recovery times, the review includes studies of recovery from brief, mild excursions of pesticide concentrations to toxic levels, as well as studies of recovery from catastrophic events (i.e., large spills). The latter are not really relevant for determination of time-to-recovery in cases of non-catastrophic events, but do provide good guidance on general aspects of ecosystem recovery.

The ability of an ecosystem to recover from a disturbance is dependent on several factors, which have been described by various authors in slightly different terms at different times. Cairns and Dickson (1977) give seven factors that determine how rapidly the recovery process will be after an ecosystem has suffered damage from one or more stressors (chemical or physical):

(1) severity and duration of the stress;
(2) number and kinds of stressors;
(3) residual effects on the physical environment (e.g., from dredging or building dams);
(4) presence of epicenters or refugia for recolonization;
(5) innate vulnerability of the system (e.g., lack of structural/functional redundancy);
(6) inertia of the system (resistance to change);
(7) resilience of the system (ability to readjust after exposure to a stressor).

Later, Cairns (1990) discussed, and quantified, six similar factors that affect an ecosystem's ability to recover from major ecological disasters: (a) existence of nearby sources of re-colonizing organisms; (b) voluntary or involuntary transportability of eggs, spores, larvae, flying adults, or other life stage; (c) condition of habitat following stress; (d) presence and persistence of residual toxicants following stress; (e) chemical-physical environmental quality following stress; and (f) potential for management/other agencies to assist in remediation. By rating each of these factors with a score of $1 = $ poor, $2 = $ moderate, or $3 = $ good, the following simple recovery index was defined:

$$\text{Recovery Index} = a \times b \times c \times d \times e \times f \qquad (10)$$

A score of 400+ indicates that the ecosystem has an excellent chance of rapid recovery, 55–399 indicates a fair to good chance of rapid recovery, and <55 indicates a poor chance. Rapid recovery is defined as having 40–60% of species reestablished within the first year after a major exposure event, and as many as 95% reestablished within 3 years. Using this system (and recognizing that excursions above water quality criteria are not in the realm of ecological disasters, but are rare, brief, and mild events of limited scope), the following scores may be assigned to ecosystems in the Sacramento and San Joaquin River basins that have had excursions above water quality criteria:

$a = 1-3$; unaffected nearby tributaries are expected to be present, except in highly urbanized or heavily agricultural areas;
$b = 1-3$; depending on degree of transportability of species in damaged community;
$c = 3$; exceedance is not expected to result in habitat alteration;
$d = 1-3$; depending on chemical;
$e = 3$; no chemical-physical alterations expected;
$f = 3$; regulatory intervention expected.

These values give a recovery index ranging from 27 to 729, indicating that the chance of rapid recovery within 3 years following a major ecological disaster is poor to excellent. It is reasonable to assume that rapid recovery (as defined by Cairns 1990) is probable in much of the Sacramento and San Joaquin River basins following brief, mild, limited-scope excursions above criteria levels. However, when all nearby tributaries are affected by the toxicant, where species living in the affected area are not readily transportable, and/or where the chemical either does not readily dissipate or is an ongoing contamination source, recovery may be hindered. In such cases, site-specific frequency components of criteria may need to be derived.

Similarly, Yount and Niemi (1990) report that the reasons behind cases where rapid recovery has been observed in lotic systems include (1) the life history characteristics of organisms allowed rapid recolonization and repopulation, (2) unaffected up- and down-stream areas were available to supply new organisms, (3) lotic systems have high rates of flushing, and (4) organisms that live in lotic systems have evolved a lot of flexibility and adaptability because lotic systems are variable environments.

Yount and Niemi (1990) also summarized studies of ecosystem recovery following application of piscicides for eradication of undesired fish. In studies of rotenone treatments, fish recovery took from 12 to 16 months and benthic macroinvertebrate recovery took 12 months (Charles 1958; Little 1966). Similarly, in California, macroinvertebrates recovered from rotenone applications within 6 months (Cook and Moore 1969), whereas in Scotland they recovered within 1 year (Morrison 1977). Macroinvertebrates had not recovered 1 year after application of toxaphene in Alaska (Meehan and Sheridan 1966). Jacobi and Degan (1977) reported full recovery of macroinvertebrates 1 year after antimycin application, and Minckley and Mihalick (1981) observed complete recovery of benthic invertebrates 3 years after antimycin application (in this case, recovery may have occurred sooner, but the investigators did not check).

Whiles and Wallace (1995) studied macroinvertebrates in headwater streams exposed to methoxychlor in four seasonal treatments over a period of 4 years. They concluded that abundance measures were not necessarily the best measure of recovery, but that ecosystem structure is important. They found that ecosystem recovery was dependent on the life cycles of the taxa making up the system. For example, organisms with shorter life cycles or extended flight capability are able to recolonize more rapidly leading to more rapid ecosystem recovery. In this study, the ecosystem required 2 years to recover nearly to its pre-treatment structure, but even then there were slight taxonomic and developmental stage differences.

In a review of studies of ecosystem recovery after damage from runoff of pesticides after forest application, Yount and Niemi (1990) reported recovery times ranging from 2 to 3 months for selected species of insects exposed to DDT (Hoffman and Drooz 1953) to more than 4 years for other insects (Hastings et al. 1961). Ephemeroptera exposed to aldrin required more than 19 months to recover, while Trichoptera and Chironomidae recovered in less than 19 months (Moye and Luckmann 1964). Arthropod biomass was reduced by exposure to fenitrothion, but recovered in 50 days (Eidt 1981). Considering fish species in situations of exposure due to runoff, Warner and Fenderson (1962) reported recovery of trout populations 3 years after DDT exposure. Keenleyside (1959) concluded that Atlantic salmon populations were able to recover if DDT application was not repeated more than once every 3 years. In contrast, when forests were sprayed with DDT during hatching and smolt migration periods, Atlantic salmon recovery required 4–6 years because of near elimination of an entire age class (Elson 1967). In areas where DDT had been sprayed repeatedly for several years, recovery took longer than 9 years. The longer recovery periods in this study were related to the relatively long (5–6-years) life cycle of Atlantic salmon and to the fact that the DDT was applied on a watershed scale resulting in isolation of the affected areas from refugia that might have provided migrants to repopulate the affected area. Such a drastic exposure from watershed-scale pesticide applications are not likely to occur in the Sacramento and San Joaquin River basins, thus the probability of eliminating, or nearly eliminating, an age-class from an entire basin is remote.

Studies of recovery times are also summarized by Yount and Niemi (1990) for cases of direct application of pesticides to water bodies. Corbet (1958) reported recovery of insect larvae in 40 days after DDT application to streams in Uganda. After treatment with methoxychlor, Chironomidae recovered in 1–2 weeks, Plecoptera in 5 weeks, and blackflies in 2–4 weeks (Fredeen 1975, 1983). Snail populations recovered after treatment with the molluscicide, Bayluscid®, after 10 months in hard water, and after 22 months in soft water (Harrison and Rattray 1966). After treatment with 3-trifluoromethyl-4-nitrophenol (TFM), benthic organisms in Lake Superior and Lake Michigan recovered in 1 year (Torblaa 1968), while Dermott and Spence (1984) observed recolonization of invertebrates in streams within 3 weeks of TFM treatment. In a follow-up experiment, Jeffrey et al. (1986) found that benthic invertebrates had not recovered more than 35 days after TFM treatment of a stream. The difference between the two studies was that the Jeffrey et al. (1986) study was conducted in colder weather when convective currents caused mixing of TFM 55 cm into the hyporheic zone. The Dermott and Spence (1984) study was conducted in warmer weather when the convective forces were not present and the uncontaminated hyporheic zone served as a refuge from TFM exposure.

A few other studies of recovery after pesticide exposure were reviewed by Yount and Niemi (1990). Ghetti and Gorbi (1985) simulated an accidental spill of parathion that would produce measurable, but not catastrophic, effects on macroinvertebrates. Their test stream recovered in 3 months with macroinvertebrate density equaling or exceeding that of a reference stream after 117 days. However, macroinvertebrate trophic structure required 2 years for recovery (Wallace et al. 1986). In Nigeria, an isolated pool was treated with the insecticide Gammalin-20 (lindane) to kill fish. After 1 month, no fish were found, but after 3 months the total number of taxa had returned to the pre-treatment state (Victor and Ogbeibu 1986).

One approach to measuring the ability of individuals or species to recover from pulse exposures to pesticides is to compare LC_{50} values obtained in continuous exposure experiments with those obtained in pulsed exposure experiments. For example, Parsons and Surgeoner (1991a) found no difference in LC_{50} values for mosquito larvae exposed to four 1-h pulses of carbaryl compared to those exposed continuously for 4 h. They concluded that this indicated that there was no recovery during the 12-h intervals between pulses.

Turning to more recent studies, Liess and Schulz (1999) observed that in streams exposed to parathion, four of 11 species of macroinvertebrates that had disappeared after treatment recovered in 6 months; nine species had recovered after 11 months; two species remained at low density for the full year of the study. Cold and Forbes (2004) found that *G. pulex* recovered in 2 weeks from a 1-h pulse exposure to esfenvalerate. Heckmann and Friberg (2005) showed that macroinvertebrate community structure had recovered within 2 weeks from two 30-min pulses of lambda-cyhalothrin. *D. magna* recovered to control levels of total neonates per female and population growth rate within 21 days after a 24-h pulse exposure to fenvalerate (Reynaldi and Liess 2005). Schulz and Liess (2001) observed chronic effects on populations of insect larvae more than 240 days after 1-h exposures to fenvalerate. Parsons and Surgeoner (1991b) reported that mosquito larvae exposed to 0.5- to 4-h pulses of permethrin, carbaryl and carbofuran were able to recover from immobility, but not after 8- to 24-h exposures.

3.4.2 Allowable Frequency of Exceedance – Conclusion

In setting an allowable frequency of exceedance of a water quality criterion, the question is how much time it would take for organisms at various organizational levels to recover from brief pulse exposures to contaminants. Yount and Niemi (1990) concluded that ecosystem recovery from pulse exposures generally occurs in less than 3 years, and often in less than 1 year. Species that are slowest to recover are those with the longest life cycles. Similarly, Niemi et al. (1990) concluded that most ecosystems are able to recover from disturbances in less than 3 years except in cases where physical habitat was altered, the system was isolated, or residual pollutant remained. The majority of reviewed studies that addressed community-, population-, or species-level effects indicated that recovery occurs in 3 years or less. The only exception is the study by Elson (1967), but, as discussed, the exposure conditions of that study were extreme and not particularly relevant to cases of brief, mild excursions above a water quality criterion.

Based on this review, 3 years between exposure events should allow full recovery from effects of an excursion above either acute or chronic water quality criteria in the Sacramento and San Joaquin River basins. This is in agreement with USEPA (1985, 2003a) methodologies, although the 3-years frequency component was supported by minimal data when it was first proposed. Acute and chronic criteria derived by the UCDM include a statement that exceedances should not occur more than once every 3 years.

4 Water Quality Effects

Because water quality criteria are derived from laboratory studies conducted in carefully controlled systems, it is necessary to consider the effects that water quality characteristics may have on the toxicity of a chemical in the environment. For pesticides, the major concerns are the effects of suspended and dissolved particulate matter on bioavailability, the effects of pesticide mixtures, and the effects of temperature, pH, or other water quality parameters on toxicity.

4.1 Bioavailability

The issue of bioavailability was discussed in TenBrook et al. (2009), but this important and complex topic merits further exploration because there is a level of uncertainty in deriving criteria with data from laboratory studies conducted in clean water (i.e., solid-free), and then using those criteria to protect aquatic organisms in an environment that contains varying levels of solids. Two questions need to be addressed. First, which phase or phases of a chemical in water are bioavailable? And, second, when bioavailability is well understood for a particular pesticide, what is the best way to determine compliance with water quality criteria? The first question is addressed through a literature review. Options for addressing the second question include equilibrium partitioning models, direct analysis of pesticides in different phases, and the use of passive sampling devices to estimate concentrations of bioavailable pesticides. Each of these is discussed below.

Staples et al. (1985) consider bioavailable chemicals to be those available to exert toxicity or to bioaccumulate. They hypothesized that most neutral organic chemicals must be in solution in order to be bioavailable, and they cite studies with polynuclear aromatic hydrocarbons (PAHs), crude oil components, and polychlorinated biphenyls that support their view (Anderson et al. 1977; Halter and Johnson 1977; Neff 1979; Roesijadi et al. 1978a, b; Rossi 1977). Similarly, DiToro et al. (1991) concluded that compounds bound to either sediments or dissolved organic carbon (DOC) are not bioavailable. Reductions in accumulation and toxicity of synthetic pyrethroids in *D. magna* have been related to the binding of these compounds to

DOC (Day 1991), and reduction in bioconcentration of chlorobenzuron in *D. magna* was attributed to binding of the compound to dissolved humic material (Steinberg et al. 1993). Kukkonen and Oikari (1991) showed that the total concentration of dissolved organic matter (DOM) in water is one of the main factors that controls bioavailability of organic contaminants. Pyrethroid uptake and toxicity in *D. magna* and *C. tentans* in water-sediment systems was reduced with increasing sediment OC concentration (Maund et al. 2002). There is a species-specific aspect to the effects of solids on bioavailability. Sediment-biota accumulation factors (BSAF) for PAHs were reduced for the marine polychaete, *Nereis diversicolor,* in sediments amended with activated carbon, but there was no change in the BSAF for the gastropod *Hinia reticulata* (Cornelissen et al. 2006).

The studies discussed so far provide evidence that solids in water affect bioavailability of organic contaminants, but they do not necessarily support the hypothesis that only freely dissolved contaminants are bioavailable. Several studies refute the "dissolved = bioavailable" concept. Dissolved humic material decreased the toxicity of diazinon, 4-chloroaniline, and 4-nitrophenol, had no effect on the toxicity of tetrabromobisphenol-A, o-toluidine, 3,4-dichloroaniline, and pentachlorophenol, but increased the toxicity of 2,4-dichlorophenol and 2,4,5-trichlorophenol to *D. magna* (Steinberg et al. 1992). Fewer walleye survived exposure to chlorpyrifos-humic acid (HA) complexes than to either HA alone or chlorpyrifos alone, and no differences were seen in cholinesterase inhibition between chlorpyrifos-HA and aqueous chlorpyrifos exposures (Phillips et al. 2003). Schnürer et al. (2006) showed that glyphosate was microbially degraded even when sorbed to soil.

Without pesticide-specific, species-specific, site-specific information regarding which phases are bioavailable, compliance must be based on measurement of total pesticide concentration in water. If bioavailability information is available for a specific case, then there are several approaches that regulators may use to determine compliance. Each is discussed below.

Case 1: Pesticide is bioavailable in all three phases or no information is available. If studies show that a pesticide is bioavailable in solid, dissolved-solid, and freely dissolved phases, or if nothing is known about bioavailability for a particular pesticide on a site-specific basis, then compliance must be determined on the basis of the total concentration of pesticide in whole water.

Case 2: Pesticide is bioavailable in fewer than three phases. In this case, regulators still have the conservative option to determine compliance based on total pesticide concentration. However, if site-specific information is available, then compliance determination may be refined by consideration of just the bioavailable fraction or pesticide in water. The most direct approach is to measure pesticides in each phase individually and then determine the total bioavailable concentration by adding together the results from each bioavailable phase. Exploring analytical options is beyond the scope of this project, but several studies have measured pesticide concentrations in three phases by various methods (e.g., Eadie et al. 1990; Liu et al. 2004; Rogers 1993). Two other options for determining compliance based on the quantity of bioavailable pesticide are discussed below. The first is a modeling approach applicable in cases where only the dissolved fraction is bioavailable.

The second is the use of passive sampling devices, which are applicable to any combination of phase bioavailability.

To address the case of bioavailability in the freely dissolved phase, Staples et al. (1985) developed a simple model to describe the relationship between total and dissolved concentrations of a contaminant in water that is based on the concentration of suspended sediment in the water and the solid–water partition coefficient. This model is the one used in RIVM (2001), for converting total concentrations to dissolved concentrations:

$$C_{\text{dissolved}} = \frac{C_{\text{total}}}{1 + (K \cdot S)} \qquad (11)$$

where

$C_{\text{dissolved}}$ = concentration of chemical in dissolved phase;
C_{total} = total concentration of chemical in water;
K = solid–water partition coefficient (L/kg), expressed as $K_{\text{OC}}/f_{\text{OC}}$;
S = concentration of sediment in water (kg/L).

One problem with this approach is that it makes no distinction between suspended solids and dissolved solids, both of which affect bioavailability (DiToro et al. 1991; McCarthy et al. 1985), but which can have very different partition coefficients (Delle Site 2001). Measured DOC–water partition coefficients (K_{DOC}) for fluoanthene were incorrectly estimated in 11 different sediment pore waters using models that assume $K_{\text{DOC}} = K_{\text{OC}}$ (Brannon et al. 1995). In a study of phenanthrene binding and sorption to HAs, Laor et al. (1998) found that partition coefficients for dissolved HA were at least one order of magnitude higher than coefficients for mineral-associated HA. Thus, trying to describe the partitioning process based only on a K_{OC} value is an oversimplification.

To improve this model, researchers have expanded it to include three phases: freely dissolved, adsorbed to DOM, and adsorbed to solids (Eadie et al. 1990; Liu et al. 2004). The Great Lakes criteria derivation methodology uses a three-phase model for derivation of human health and wildlife criteria (Eadie et al. 1990; USEPA 2003a). The three-phase model is an improvement over the two-phase model, but it does not acknowledge that partition coefficients vary considerably depending on the nature of the solids. Normalizing the partition coefficient to OC is a common approach to reduce that variability, but even K_{OC} values vary depending on the nature of the OC. Delle Site (2001) notes that the sorptivity of organic matter depends on the relative proportion of humic and fulvic acids, lipids, and humins (based on studies by Chiou et al. 1986, 1987; Garbarini and Lion 1986; Gauthier et al. 1987). Kukkonen and Oikari (1991) reported that the degree of aromaticity and portion of hydrophobic acids in DOM are important controlling factors in the sorption of organic compounds. The log K_{OC} values for 4-nonylphenol were 1.71, 3.08, 4.15, 4.50, and 4.71 for cellulose, chitin, lignin, humic acid, and natural sediment, respectively (Burgess et al. 2005). In a study of chlorpyrifos binding to colloidal materials, Wu and Laird (2004) found that chlorpyrifos sorbed strongly to a calcium-humate and did not desorb, but moderately sorbed to and desorbed from a river sediment. They concluded that both the organic and inorganic materials in suspended sediment affect the adsorption and desorption of chlorpyrifos. In view of these studies, simple partitioning models are not useful for making general predictions about phase distributions for organic contaminants. It may be possible to use them in site-specific situations in which partition coefficients are available for the specific types of solids present in the site water.

Simple partitioning models also assume that the mechanism for reduced bioavailability and toxicity of organic contaminants, in the presence of solids, is adsorption or binding of the contaminant to the solids. However, Steinberg et al. (1992) found that dissolved humic material and sunlight enhanced diazinon degradation, which is another mechanism that reduces toxicity. This is not surprising in that photolysis of DOM is a known source of hydroxyl radicals in waters (Takahashi et al. 1988).

Semi-permeable membrane devices (SPMDs) are passive sampling devices that are intended to mimic uptake of bioavailable contaminants (Huckins et al. 1990). Using laboratory-determined uptake rate constants, SPMDs can be used to determine time-integrated ambient water concentrations of bioavailable contaminants. These devices work well both qualitatively and quantitatively for hydrophobic organic chemicals (Huckins et al. 1990; Lu and Wang 2003), but do not give reliable quantitative results for polar organics (Alvarez et al. 2004). Using SPMDs for quantitative analysis of ambient waters is problematic because each device has a specific membrane permeability, and uptake rates are affected by flow rates, temperatures, and fouling (Huckins et al. 1990). For measuring hydrophobic compounds, performance reference compounds (PRCs) have been developed to counter these variables (Huckins et al. 2002), but attempts to develop PRCs for polar organics have not been successful (Alvarez et al. 2004).

SPMDs are promoted for their ability to take time-integrated water samples such that contaminant pulses can be detected. However, this type of sampling is not useful for determination of short-term variability in ambient water contaminant concentrations (Gustafson and Dickhut 1997; Prest et al. 1998), and thus would be of little value in measuring compliance with acute criteria that must be met on the basis of 1-h average concentrations. Other problems with SPMDs include the fact that compounds are subject to photolysis, if devices are deployed in shallow water or near the surface (USGS 2000), and that they are not accurate measures of bioaccumulation because devices do not mimic biological processes (i.e., metabolism, excretion, chemical movement, feeding) that affect equilibrium concentrations in organisms (Huckins et al. 2004).

There is much evidence that solids in natural waters affect the uptake and toxicity of organic contaminants. TenBrook et al. (2009) pointed out that pesticide loadings to surface waters typically result from storm or agricultural runoff, and because suspended solids are also higher than normal during runoff events, it would be ideal if criteria could be expressed, or compliance could be determined, in terms that reflect this covariance of pesticides and suspended solids.

As the discussion in this section has shown, there is no simple way to incorporate the effects of solids on bioavailability into either criteria derivation or compliance determinations. It is not correct to assume that freely dissolved compounds are equivalent to bioavailable compounds. Further, it is too simplistic to assume that general partition coefficients are valid for all kinds of solids. Thus, there is no general way to predict bioavailability from physical–chemical parameters and water quality data. However, if studies are available that show which fraction, or fractions, of a particular pesticide is (are) bioavailable, then it may be of interest to directly measure the pesticide concentrations in those fractions to determine compliance. For pesticides that are only bioavailable in the freely dissolved phase, and for which K_{OC} and K_{DOC} values are available, use of a three-phase partitioning model is an option for translating measured total pesticide concentrations into dissolved concentrations. If technical limitations of passive sampling devices can be overcome, they offer another option for estimating bioavailable pesticide in water samples, but only for determination of compliance with chronic criteria. The UCDM includes brief guidance regarding how to address bioavailability.

4.2 Mixtures

Various approaches to addressing toxicity of mixtures were discussed in TenBrook et al. (2009). Lydy et al. (2004) also provides a good review of pesticide mixtures. They both concluded that there really is no way to derive criteria for all of the potential mixtures of pesticides that could occur in a water body. Nonetheless, there are many models available for determination of mixture toxicity. The question is whether such models, which are designed to predict toxicity of a mixture to a single species, can be adapted for use in determination of compliance to a water quality criterion that applies to multiple species. The CVRWQCB (2004) has done this by substituting water quality criteria for LC/EC_{50} values in a simple concentrations addition model and using the results to assess compliance. Felsot (2005) has shown that this is also possible with a toxic equivalence model.

How to best model chemical mixtures will depend on the nature of the mixture. All models must be applicable to mixtures of two or more components. Where models will differ is in whether they apply to chemicals of similar or different modes of action, and whether the mixtures show additive toxicity, or if there are known interactions leading to antagonistic (less than additive) or synergistic (greater than additive) effects. Pesticide modes of action are often known; pesticide interactions are not as well-studied. The following discussion is broken down according to the mixture models that are available. Each is described and discussed in terms of its applicability to particular types of mixtures or to determination of compliance with water quality criteria. The SSD approach to mixtures (Traas et al. 2002) was explored in TenBrook et al. (2009), but is not discussed further here as it is fairly complex, it is not readily adaptable for use in compliance determination, and it requires that SSDs be available for each component of the mixture, which would limit its applicability.

4.2.1 Additivity

There are two basic models to describe additive mixture toxicity. The concentration addition model is used for chemicals that have similar modes of action, while the independent action model is used for chemicals with different modes of action (Plackett and Hewlett 1952).

A. Concentration Addition – For Similar Modes of Action

The basic concentration addition model is expressed as (Olmstead and LeBlanc 2005; PapeLindstrom and Lydy 1997)

$$\sum_n^{i=1} \frac{C_i}{ECx_i} = 1 \text{ TU} \tag{12}$$

where

C_i = concentration of the ith chemical in the mixture;
ECx_i = concentration of the ith chemical that elicits the same response (x) as the full mixture;
TU = toxic unit.

If the sum is 1 TU, then the mixture exhibits additivity. If the sum is greater than 1 TU, the mixture is less than additive (antagonistic); if less than 1 TU, the mixture is greater than additive (synergistic). For example, if the EC_{50} is the chosen level of effect, then it is expected that the sum of the concentration:EC_{50} ratios for each component of the mixture will equal 1 TU. If less than 1 TU is required to elicit 50% effect in the mixture, then the mixture is showing more than additivity (synergism); if more than 1 TU is required, then the mixture is showing less than additivity (antagonism).

The CVRWQCB (2004) adapts this equation for determination of compliance with water quality criteria:

$$\sum_{i=1}^n \frac{C_i}{O_i} < 1.0 \tag{13}$$

where

C_i = concentration of toxicant i in water;
O_i = water quality objective/criterion for toxicant i.

As long as the sum is <1.0, the water body is considered to be in compliance with respect to the mixture.

Felsot (2005) argued that actual toxicity values (e.g., LC_{50}) should be used in these equations instead of water quality objectives for criteria, to more accurately reflect the toxicity of the mixture. This argument was made in regards to acute criteria that include a safety factor of 2 after determination of an acute value. However, for determination of compliance, the question is simply whether concentrations of chemicals in a water body are below criteria levels. To answer this question, using water quality criteria in these equations is appropriate.

As an alternative to Equation (13), Felsot (2005) suggested using the relative potency factor (RPF) approach to determine compliance in cases of additive toxicity for compounds with similar modes of action. The RPF approach is analogous to the toxic equivalency factor (TEF) approach used in assessing toxicity of dioxin and dioxin-like compounds (Van den Berg et al. 1998). By the RPF approach, one chemical (usually the most toxic) is chosen to be the reference chemical and the potency of all other similarly acting chemicals is expressed relative to the reference. The potency of each chemical is divided by the potency of the reference chemical and this ratio, the RPF, is multiplied by measured concentrations of each non-reference chemical to produce concentrations in terms of equivalents of the reference chemical. Compliance with the objective for the reference chemical is based on the sum of the measured concentration of the reference chemical plus the concentrations of the equivalents. Mathematically, this is expressed as

$$\text{TE}_{\text{total}} = C_R + \sum_{n}^{i} \text{TE}_i \qquad (14)$$

where

TE_{total} = total toxic equivalents of mixture (μg/L);
C_R = concentration of reference chemical (μg/L);
TE_i = toxic equivalents of ith component of the mixture (μg/L).

and

$$\text{TE}_i = \text{RPF}_i * C_i \qquad (15)$$

where

RPF_i = relative potency factor of the ith component of the mixture;
C_i = concentration of the ith component of the mixture (μg/L).

and

$$\text{RPF}_i = \frac{\text{EC}_{xR}}{\text{EC}_{xi}} \qquad (16)$$

where

EC_{xR} = concentration of reference chemical causing $x\%$ effect when tested alone (μg/L);
EC_{xi} = concentration of mixture component i causing $x\%$ effect when tested alone (μg/L).

For compliance determination, a multispecies RPF is needed, so Equation (16) can be written:

$$RPF_i = \frac{Criterion_{xR}}{Criterion_{xi}} \tag{17}$$

where

$Criterion_{xR}$ = water quality criterion of reference chemical (μg/L);
$Criterion_{xi}$ = water quality criterion of the ith chemical (μg/L).

Using these equations, if $TE_{total} \leq$ the criterion for the reference compound, then the water body is in compliance.

Both the concentration addition approach and the RPF/TEF approach are valid for determination of compliance in cases of additive toxicity when chemicals in the mixture have similar modes of action. Both are included in the UCDM allowing regulators to select which one works best for them.

B. Response Addition – For Independent Modes of Action

For chemicals that do not have similar modes of action, the response addition, or independent action, model is used. It is expressed mathematically as (Belden and Lydy 2006)

$$R_{mix} = 1 - \prod_{n}^{I=1}(1 - R_I) \tag{18}$$

where

R_{mix} = response of the mixture (i.e., percent response);
R_I = response expected from the Ith component the mixture.

This model is not applicable for determination of compliance. To illustrate, consider an example in which two pesticides are each present at concentrations equal to its own water quality criterion. Since the criteria are based on a concentration that might affect 5% of species, the R_I value for each pesticide is 0.05 and the equation is

$$R_{mix} = 1 - (1 - 0.05)^*(1 - 0.05) = 0.10, \text{ or } 10\% \tag{19}$$

Thus, the mixture could harm as many as 10% of species, which is unacceptable. To ensure compliance, each of the two chemicals would have to be present at a level that might be expected to harm 2.5% of species, so that R_{mix} would be at or below 5%. Estimates of percentile values below the 5th percentile are highly variable (TenBrook et al. 2009), and yet all values of interest for application of the independent action model to compliance determination for mixtures lie below this level (values above the 5th percentile would indicate non-compliance with individual criteria). While the independent action model works well for predicting toxicity of mixtures when individual toxicities are known, it is not adaptable for compliance determination.

4.2.2 Non-additivity: Synergism and Antagonism

Chemical mixtures may display non-additive toxicity in the form of either antagonistic or synergistic effects. This indicates an interaction between chemicals such that the response observed for a mixture is either less than (antagonism) or greater than (synergism) that predicted by additivity models. The concept of synergy is often used to refer to cases in which one chemical, present at non-toxic concentrations, increases the toxicity of a second chemical, but it can also be applied to mixtures in which both chemicals are at toxic levels. Mu and LeBlanc (2004) utilized the coefficient of interaction (K) to define this relationship. First described by Finney (1942), the basic equation is

$$K_x = \frac{EC50_0}{EC50_x} \qquad (20)$$

where

K_x = coefficient of interaction at synergist/antagonist concentration x;
$EC50_0 = EC_{50}$ of chemical in absence of synergist/antagonist;
$EC50_x = EC_{50}$ of chemical in presence of synergist/antagonist at concentration x.

When a measured concentration of a chemical is multiplied by the K_x value for a given concentration of a synergist/antagonist, the value that results is an adjusted, or effective, concentration of the chemical. Mathematically, this is expressed as

$$C_a = C_m(K) \qquad (21)$$

where

C_a = adjusted, or effective, concentration of chemical;
C_m = concentration measured;
K = coefficient of interaction.

For application to compliance determination, Equation (21) could be used and the effective concentration compared to the water quality criterion. Additionally, the effective concentration could be used in additivity models described in Section 4.2.1A. The difficulty is in determination of an appropriate K value. Rider and LeBlanc (2005) fit logistic functions to describe the relationship between K values and piperonyl butoxide (PBO) concentration, such that K values could be estimated for a wide range of PBO concentrations. Unfortunately, K values derived in that manner are not generally applicable. In the case of Rider and LeBlanc (2005), they are specific to the interaction of PBO with either malathion or parathion and the toxicity of binary or ternary mixtures of those chemicals to *D. magna*.

Equation (21) can be modified, in theory, for mixtures containing both synergists and antagonists, or multiple synergists/antagonists (LeBlanc, personal communication 2006):

$$C_a = C_m(K_1 K_2 ... K_n) \qquad (22)$$

where

C_a and C_m are as defined in Equation (21);
$K_1, K_2, K_n = K$ values for synergist/antagonist 1, 2...n.

Dr. LeBlanc cautions that as more K values are strung together, the error of each term will lead to large error in the adjusted concentration. Thus, this approach should not be used for compliance determination, but may be used to assess research needs.

To use the interaction coefficient concept in determination of water quality criteria compliance would require the establishment of relationships between K values and synergist/antagonist concentrations. That is, for pesticides that commonly occur together, it might be worth the research effort to establish relationships (i.e., predictive equations) between K and concentrations of known synergists/antagonists for a range of species. This issue is discussed further at the end of Section 4.2.3.

4.2.3 Combined Models

Each of the models discussed so far apply to only one type of mixture effect. In the environment, it would not be unusual to find complex mixtures that include chemicals that show all three of the basic mixture effects: additivity with similar modes of action, additivity with different modes of action, and interaction leading to synergism or antagonism. Olmstead and LeBlanc (2005) developed a model that combines concentration addition and response addition models into one. Rider and LeBlanc (2005) expanded on that model to include an interaction component. The basic model equation is

$$R = 1 - \prod_{I=1}^{N} \left\{ 1 - \cfrac{1}{1 + \cfrac{1}{\left(\sum_{i=1}^{n} \frac{k_{a,i}(C_a) \times C_i}{EC50_i}\right)^{\rho'}}} \right\} \qquad (23)$$

where

R = response of the mixture (percent of individuals responding);
N = number of cassettes (cassette = group of chemicals of similar mode of action);
I = Ith cassette;
n = number of chemicals;
i = ith chemical;
$k_{a,i}$ = interaction coefficient for chemical a (synergist/antagonist) interacting with chemical i;
C_a = concentration of chemical a in the mixture;
C_i = concentration of chemical i in the mixture;
$EC50_i$ = EC_{50} for chemical i alone;
ρ' = average power (slope) of dose-response curves of chemicals in cassette I.

This model integrates all aspects of mixture toxicity, but can only be applied to one species at a time. Gerald LeBlanc was contacted to discuss the possibility of adapting this model for determination of criteria compliance (personal communication 2006). The proposal was put forth that some of the variables in Equation (23) be redefined as follows:

R = response to the mixture (percent of species responding);
$k_{a,i}$ = multispecies interaction coefficient;
$EC50_i$ = replace with water quality criterion;
ρ' = average slope of SSDs in cassette I.

Using the model in this way, when $R > 0.05$, the criterion would not be met, because the criterion is based on the 5th percentile of the SSD. Similarly, for $R \leq 0.05$, the mixture would be in compliance.

LeBlanc's response was that this approach seemed reasonable in all respects except that it may not be possible to derive a reliable multispecies K value. His concern was that since K is a mechanism-dependent value, assuming a common value across species would be equivalent to assuming similar toxicity mechanisms across species. He gave the example of a chemical mixture containing an androgen receptor antagonist and an androgen synthesis inhibitor. In vertebrates, these chemicals interact synergistically, but in invertebrates they do not because invertebrates lack androgen receptors. Thus it may not be possible to derive a valid ecosystem K value for this mixture. Before it can be used for compliance assessment, the Rider and LeBlanc (2005) model would have to be validated for use across species. However, K values for individual species could be used to assess the potential harm from non-additive toxicity on a species-by-species basis.

4.2.4 Conclusions on Mixtures

Among the mixture approaches presented, only the concentration addition models are readily applicable for determination of compliance with water quality criteria. The interaction coefficient concept could be used if further research provided K values applicable to multiple species over a range of synergist/antagonist concentrations. The combination model of Rider and LeBlanc (2005) that incorporates concentration addition, response addition, and interaction components holds promise as a way to assess toxicity of complex mixtures, but further research and testing is needed to find a way to utilize it in compliance determination.

The UCDM includes both of the concentration addition approaches discussed in Section 4.2.1A. The non-additive model (Section 4.2.2) is also included, with the caveat that it can only be used if reliable K values are available (either a multispecies value, or individual species values). The response addition and combined models are not incorporated into the UCDM. Regulators can choose among the models and apply them to determine compliance with water quality criteria or to assess the potential for harm due to non-additive toxicity.

4.3 Other Water Quality Effects

As described by USEPA (1985, 2003a), if data are available to establish quantitative relationships between water quality characteristics and toxicity, then criteria should be expressed as equations reflecting that relationship. Both USEPA methodologies (1985, 2003a) provide detailed instructions for an acceptable method for determination of acute and chronic criteria in cases where toxicity to two or more species is related to a water quality characteristic (hardness, pH, temperature, etc.). The USEPA approach is included in the UCDM. Details are provided in Section 9.5.3.

5 Check Criteria Against Ecotoxicity Data

Once derived according to methods discussed in the procedures in Section 3, criteria must be evaluated to ensure that they are set at levels that will protect against adverse effects to (1) particularly sensitive species, (2) species within ecosystems, and (3) threatened and endangered species (TES).

5.1 Sensitive Species

Derived criteria should be compared to studies of the most sensitive species in the data set to ensure that these species will be protected. If a calculated criterion is higher than toxicity values reported for a particularly sensitive species, then the criterion may require downward adjustment, for example, by using the lower 95% confidence interval estimate of the 5th percentile, rather than the median.

5.2 Ecosystem and Other Studies

As recommended in Section 2.1.4, criteria should be evaluated against field or semi-field data to judge whether they will be protective of all species within ecosystems. If not, then criteria may need to be adjusted downward. This is consistent with several criteria derivation methodologies (OECD 1995a; RIVM 2001; USEPA 1985, 2003a; Zabel and Cole 1999) and is included in the UCDM.

5.3 Threatened and Endangered Species

A number of TES live in the waters of California's Central Valley. Because TES are protected, it is likely that very little toxicity test data will be available for them. Nonetheless, it is important to ensure that they are protected by water quality criteria. Certainly, if data of acceptable quality are available for TES, then those data should be included in sets used for criteria derivation. Derived criteria should be checked against toxicity values for TES to ensure that criteria will protect them. Because criteria were calculated to be estimates of ecosystem no-effect levels, they should be protective of TES. However, it is worthwhile to use available data and tools to confirm this.

When effects data are lacking for TES, the content of both the USEPA (2003a) and the Australia/New Zealand (ANZECC and ARMCANZ 2000) guidelines suggest that studies with appropriate surrogate species may be used to set criteria. Deciding on appropriate surrogate species is left to professional judgment. More rigorous methods for estimating toxicity to TES, based on surrogates, are provided by both QSARs and ICEs. The ICE model (Asfaw et al. 2003) is available for free download at http://www.epa.gov/ceampubl/fchain/index.htm. It can be used to estimate the toxicity of any chemical for which a toxicity value is available for a surrogate species; thus far, the ICE model has only been developed to render estimates of acute toxicity. Moreover, the ICE model works best when making correlations within families. It can be used for larger taxonomic distances, but estimates will not be as good. QSARs are available for both acute and chronic toxicity (e.g., RIVM 2001), but, as discussed in Section 2.2.1, application of QSARs is not well developed for chemicals with non-narcotic modes of action; this severely limits QSAR usefulness in estimating pesticide toxicity.

To assess whether criteria derived by the UCDM will be protective of TES (based on the most recent list available), the following procedure is proposed.

For comparison to acute criteria:

(1) Compare criteria to toxicity values from acceptable studies of effects on TES.
(2) If no toxicity values are available for a TES, but an acceptable acute toxicity value is available for a surrogate species in the same family or genus as the TES, then use the ICE (v. 1.0) program to estimate a toxicity value for the TES; compare this estimated value to the acute criterion.
(3) If no surrogate value is available, and if the chemical of interest has a narcotic mode of action, select a QSAR (e.g., from OECD 1995a; RIVM 2001) that can be used to estimate toxicity based on a log K_{OW} value.

For comparison to chronic criteria:

(1) Compare criteria to toxicity values from acceptable studies of effects on TES.
(2) If the chemical of interest has a narcotic mode of action, select QSARs (e.g., from OECD 1995a; RIVM 2001) that can be used to estimate toxicity based on a log K_{OW} value.

If no data for the TES or acceptable surrogates are available, or if QSARs are not applicable, then it will not be possible to assess whether changes in the criteria are required to be protective of these species. If any of the above comparisons reveal that a criterion is higher than any of the TES toxicity values (or estimated toxicity values), then the criterion may need to be adjusted downward.

6 Partitioning to Other Environmental Compartments

Although partitioning of a pesticide to other environmental compartments is of concern to environmental managers, it is beyond the scope of water quality criteria. This section is intended to guide the evaluation of the agreement between the water quality criteria and existing guidelines for wildlife, humans, air, and sediment. If these criteria are found to be in conflict with any existing guidelines, this fact should be flagged for review by environmental managers, but the criteria should not be adjusted.

6.1 Bioaccumulation/Secondary Poisoning

The goal of the UCDM is to establish water quality criteria that protect aquatic life, which are not directly concerned with the protection of terrestrial wildlife or human health. However, for potentially bioaccumulative chemicals it is important to be sure that water quality criteria are set at levels that do not lead to unacceptable levels of chemicals in food items. In the UCDM we include a procedure for checking calculated chronic criteria for the possibility of secondary poisoning of wildlife, or possible human health effects, which results from bioaccumulation in fish or other food items. Acute criteria do not require this check because they are intended to protect against short periods of elevated pesticide concentrations, rendering the equilibrium model inappropriate. For wildlife, this requires the availability of studies that demonstrate adverse effects from dietary intake of toxicants; for human health, this requires the availability of US Food and Drug Administration (USFDA) action limits for the chemical of concern. The procedure described herein is based on the content in Section 7.3.2 of TenBrook et al. (2009). The discussion is framed in terms of fish tissue, but the procedure may be applied to any potential food items.

The first step in the process is to determine if the chemical of interest has the potential to bioaccumulate. The OECD (1995a) provides useful guidance on this point, which is incorporated into the UCDM. According to this guidance, chemicals are likely to bioaccumulate if they have a log K_{OW} >3, molecular weight <1,000, molecular diameter <5.5 Å, and molecular length <5.5 nm. The latter two parameters are not readily available for many chemicals, but may be used as guidelines if available. Chemicals are not expected to bioaccumulate if they are reactive and/or readily metabolized.

The next steps only apply if a chemical is determined to have bioaccumulative potential, and if dietary toxicity data or USFDA action levels are available. Based on an equation in the EU risk assessment technical guidance document (ECB 2003), TenBrook et al. (2009) proposed the following equation for translating dietary NOEC or LC_{50} values or USFDA action levels to water NOEC values:

$$\text{NOEC}_{\text{water}} = \frac{\text{NOEC}_{\text{oral\_predator}}}{\text{BCF}_{\text{fish}} \times \text{BMF}_{\text{fish}}} \quad (24)$$

or

$$\text{NOEC}_{\text{water}} = \frac{\text{LC}_{50,\text{oral\_predator}}}{\text{BCF}_{\text{fish}} \times \text{BMF}_{\text{fish}}} \quad (25)$$

where

$\text{NOEC}_{\text{water}}$ = NOEC in water; concentration in water below this level is not expected to lead to bioaccumulation to harmful levels in food items;
$\text{NOEC}_{\text{oral\_predator}}$ = dietary NOEC value for wildlife or USFDA action level;
$\text{LC}_{50,\text{oral\_predator}}$ = dietary LC_{50} value for wildlife (mg pesticide/kg food);
BCF_{fish} = bioconcentration factor; ratio of concentration of chemical in tissue due to water-only exposure to concentration in water; whole-body, wet wt value (ECB 2003; OECD 1995a);

Table 22 Default BMF values (ECB 2003)

log K_{OW}	BCF	BMF
<4.5	<2,000	1
4.5–<5	2,000–5,000	2
5–8	5,000	10
>8–9	2,000–5,000	3
>9	<2,000	1

BMF_{fish} = biomagnification factor in food item; ratio of concentration of chemical in predator to concentration in prey items; based on lipid-normalized values, if available (ECB 2003).

If no measured BMF is available, then an appropriate default value (Table 22; based on log K_{OW} or BCF) should be used (ECB 2003). Alternatively, if a bioaccumulation factor (BAF) is available for fish, then Equation (24) may be modified to

$$NOEC_{water} = \frac{NOEC_{oral\_predator}}{BAF_{fish}} \quad (26)$$

where

$NOEC_{water}$ = NOEC in water;
$NOEC_{oral\_predator}$ = dietary NOEC for wildlife or USFDA action level (mg pesticide/kg food);
BAF_{fish} = bioaccumulation factor in fish; ratio of concentration of chemical in tissue due to water plus dietary exposure to concentration in water.

In this form, the equation is similar to the USEPA (USEPA 1985, 2003a) method for determination of a final residue value, but for the UCDM, the BAF may not simply be replaced with a BCF value. If no BAF value is available, then Equation (24) or Equation (25) must be used, and if no measured BMF value is available, then the appropriate default value should be used (Table 22). If multiple BCF, BAF, or BMF values are available for a chemical, the geometric mean of all acceptable values should be used.

To determine compliance, the $NOEC_{water}$ derived from either Equation (24), or (25), or (26) would be compared to the chronic water quality criterion. If it exceeds the criterion, then no adjustment of the criterion is necessary. If the $NOEC_{water}$ is below the criterion, then this should be indicated in the final criteria statement that these criteria may not be protective of all beneficial uses, based on the bioaccumulation/secondary poisoning section and that additional review is needed.

6.2 Harmonization with Sediment and Air Criteria

The final element that regulators may wish to consider is whether a pesticide that is present in the water at criteria levels might have the potential to move from the water compartment into another environmental compartment (i.e., sediment, biota, air), where it may exceed levels of concern established for those compartments. This is the concept of harmonization and was discussed in TenBrook et al. (2009). Consideration of bioaccumulation and secondary poisoning is a specific case of harmonization of criteria between two compartments. Because this is an assessment of equilibrium conditions, it is only necessary to consider chronic criteria concentrations, as we did for bioaccumulation (Section 6.1). This procedure is not necessary if levels of concern have not been established, and toxicity data are not available, for air, sediment, or biota compartments.

Steady-state environmental models are tools that may be used to assess harmony, or coherence, across environmental media. Acceptable models are available for free over the internet. The Exposure Analysis Modeling System (EXAMS; Burns 2004) is available from the USEPA Center for Exposure Assessment Modeling (CEAM; http://www.epa.gov/ceampubl/swater/index.htm). Mackay's (2001) Fugacity-Based Environmental Equilibrium Partitioning Models Levels I, II, and III are available as free downloads from the Canadian Environmental Monitoring Center (CEMC; http://www.trentu.ca/cemc/welcome.html). To estimate equilibrium concentrations of pesticides in various compartments at equilibrium, the easiest-to-use and most appropriate model is the Level I fugacity model. The EXAMS and Level II and III fugacity models can be used, but are more complex and require more site-specific input. The Level I fugacity model requires several input parameters including water solubility, vapor pressure, melting point, and log K_{OW} for prediction of steady-state concentrations of chemicals in air, water, suspended sediment, bed sediment, and biota. Steady-state environmental models allow one to determine equilibrium concentrations in various compartments. The equilibrium that exists between any two compartments may be described by the following equation (based on a simple first-order kinetic model):

$$C_1 k_{12} = C_2 k_{21} \qquad (27)$$

where

C_1 = concentration of pesticide in compartment 1;
C_2 = concentration of pesticide in compartment 2;
k_{12} = rate constant for transfer of pesticide from compartment 1 to compartment 2;
k_{21} = rate constant for transfer of pesticide from compartment 2 to compartment 1.

Equation (27) can be rearranged:

$$C_1/C_2 = k_{21}/k_{12} = K \qquad (28)$$

where

$K =$ the equilibrium constant between the two compartments.

In the simulations that will be run to assess harmonization in the UCDM, the concentration of pesticide in water will be set at the chronic criterion level by adjusting the total mass of pesticide in the system. According to Equation (28), as long as C_1 is constant, C_2 will also be constant for a given equilibrium constant. Thus, the only kinds of model input changes that will affect final concentrations in non-water compartments are those that affect the equilibrium constant. For example, changing lipid levels in fish, or the OC content in suspended sediments will cause changes in equilibria, but changing concentrations of solids, or volumes of air or water will not. Model simulations can be run over a range of values to provide information applicable to a variety of site-specific situations.

Model outputs, based on having a chemical of concern at its chronic criterion level in water, should be compared to appropriate levels of concern (if any) established for the non-water compartments (e.g., sediment or air quality criteria or USFDA action levels). If the steady-state concentrations in all compartments are below their respective levels of concern, then the water quality criterion is acceptable. If not, then it should be indicated in the final criteria statement that these criteria may not be protective of all beneficial uses as a result of the harmonization procedure and that additional review is needed. However, before any adjustment is made, further site-specific modeling is recommended. The UCDM includes a harmonization procedure and a list of available, acceptable models with documentation.

7 Assumptions and Limitations

The assumptions, limitations, and uncertainties involved in criteria generation should be included in the criteria methodology and associated reports so that environmental managers are aware of the accuracy and confidence in the criteria. This review paper was designed to present many of those assumptions and limitations. Major assumptions, limitations, and uncertainties are reviewed here, followed by recommendations for the improvement of the criteria derivation process.

7.1 Assumptions, Limitations, and Uncertainties of the UCDM

One benefit of calculating criteria using a statistical distribution is that it provides a quantitative measure of the uncertainty in the estimate. Several calculations of distributional estimates will be included in criteria reports (Section 3.1.3), and the uncertainty in the resulting criteria can be seen by comparing the median estimate to the estimate with 95% confidence. In the UCDM, the confidence level of these estimates is calculated assuming that the uncertainty in the fitted distribution is the greatest source of uncertainty in the criteria calculation. The variation in species sensitivities is likely to outweigh other calculable sources of uncertainty (Section 3.1.3), such as the uncertainty of an LC_{50}, which could be expressed as the confidence limits of the reported LC_{50}.

However, these confidence intervals do not include uncertainty from the assumptions listed in Section 3.1.5A, which are universal to any method that employs SSDs for protection of all species. These may be the most important assumptions influencing the effectiveness of criteria; unfortunately, the uncertainty associated with these assumptions is nearly impossible to quantify at this time. Moreover, although the UCDM was designed to incorporate many considerations that are not yet in many methodologies, it is impossible to account for them all. Some, such as sublethal effects and additive effects with other compounds (i.e., additive effects of compounds with different modes of action), are too complex to incorporate here. Such issues and others could lead to underprotection, whereas others could lead to overprotection. Further, the models included will probably often be limited by available data.

Few chronic data are often available, which will be limiting for any method. The UCDM employs ACRs, which are a fairly common means to provide protective estimates when chronic data are lacking. ACRs are based in the assumption that it is possible to extrapolate from toxicity data at one life stage to an entire life cycle. When sufficient pesticide-specific data are not available, the UCDM also provides a default ACR and default AFs. The AFs were derived empirically using only pesticide data because the UCDM is not concerned with other types of contaminants. However, the AFs are limited by the relatively small amount and diversity of pesticide data available.

Other limitations are likely to be encountered when deriving a particular criterion, and these should be addressed in individual criteria reports. Final criteria statements should briefly address any data limitations that affect the procedure used to determine the final criteria. The thrust is to make it obvious how the final criterion was derived. An example of an important limitation affecting the derivation process would be missing taxa that required use of AFs instead of an SSD. Any other considerations (Sections 4, 5, and 6) should be included that may be important for policy makers to consider.

7.2 Data Generation to Improve Criteria Derivation

In the process of creating the UCDM, specific areas were identified in which more data would help to derive better criteria. This section highlights some of the procedures most in need of additional data and research.

For many pesticides, one of the largest limiting factors in the certainty of calculated criteria is the amount of data available, especially for newer compounds.

Although a large amount of toxicity data may be generated for pesticide registration, a pesticide may be registered with as little as three acute toxicity studies on freshwater organisms: one invertebrate and one or two fish. Although, in practice, data on more than three species are generated, it is uncommon to have a data set sufficiently large to allow use of established methods such as USEPA (1985). Chronic data are especially needed because pesticide regulation requires only one fish early-life stage test and one invertebrate life-cycle test for freshwater chronic studies (US Code Title 40 2009). It would also be helpful if chronic studies were reported with both hypothesis testing and regression analysis (EC_x) because regression methods are now often preferred, but most existing data are from hypothesis tests (Section 2.1.2).

It will be impossible to obtain large data sets for all compounds, making it necessary to have some means to set limits for pesticides that have only minimal data. The ACR and AF procedures, derived from richer pesticide data sets, may be used as estimates for compounds without much data. However, most of the richer pesticide data sets available now do not reflect many of compounds currently in use. The default ACR would benefit from the generation and incorporation of more multispecies pesticide ACRs, making the default ACR a better representative of currently used pesticides. Similarly, the AFs would benefit from the addition of more acute pesticide data sets. Sections 3.2.5C and 3.2.4 include the procedures used to derive ACRs and AFs, respectively, so that they may be recalculated when more data become available.

Given the frequent data limitations, criteria reports would benefit from periodic review and incorporation of data from the most recently published literature.

8 Guideline Format

Water quality criteria methodology must be understandable, navigable, and usable by environmental managers (TenBrook et al. 2009). The UCDM is presented below in Section 9 and includes a flow chart designed to outline the derivation procedure; in addition, it includes explicit guidance and instructions for each step of the process, details of calculations, and numerous tables of information. For clarity, background and supporting information have been separated from the UCDM itself. The rationale for selecting various components and approaches is discussed by TenBrook et al. (2009), and in this review as well.

9 The UCD Methodology

9.1 Goals and Definitions

The goal of the UCDM is to extrapolate from available pesticide toxicity data for a limited number of species to a concentration (criteria) that should not produce detrimental physiological effects in aquatic life. These criteria aim to protect all species in the ecosystem. The UCDM was designed for the Sacramento and San Joaquin River watersheds, but is generally applicable to freshwater ecosystems in the United States. Simple modifications could be made to adapt this method for saltwater criteria or other geographic areas.

9.1.1 Relevant Compounds

This method is intended for deriving water quality criteria for pesticides. The term pesticide is defined by CVRWQCB (2004) as (1) any substance or mixture of substances that is intended to be used for defoliating plants, regulating plant growth, or for preventing, destroying, repelling, or mitigating any pest, which may infest or be detrimental to vegetation, man, animals, or households, or be present in any agricultural or non-agricultural environment whatsoever, or (2) any spray adjuvant, or (3) any breakdown products of these materials that threaten beneficial uses. Certain procedures were derived using only data on organic pesticides and have not been validated for metals or other inorganic compounds. This is noted in the AF Section (9.3.3) and in the default ACR Section (9.4.2C).

9.1.2 Definition of Numeric Criteria

Water quality criteria are referred to by different terms and are used for different purposes depending upon how they are derived (TenBrook et al. 2009). For the UCDM, numeric criteria are science-based values that are intended to protect aquatic life from adverse effects of pesticides, without consideration of defined water body uses, societal values, economics, or other non-scientific considerations.

9.1.3 Overview

The UCDM consists of a combination of features from existing methodologies that have been refined, based on recent research in aquatic ecotoxicology and environmental risk assessment. Components of the UCDM were selected and based on evaluations and recommendations in TenBrook et al. (2009), and in Sections 1, 2, 3, 4, 5, 6, 7, and 8 of this review. The UCDM includes components for deriving water quality criteria from both large and small data sets. For a given compound, the criteria derivation procedure selected depends on the richness of the available data. Figures 3 and 4 are flow-charts that summarize procedures for collecting, evaluating, and reducing data sets, and for deriving acute and chronic criteria. The UCDM is herein presented in the format of a standard operating procedure.

9.2 Data

This section provides details of how to collect, summarize, evaluate, and reduce data to be used in criteria derivation.

The University of California-Davis Methodology for Deriving Aquatic Criteria

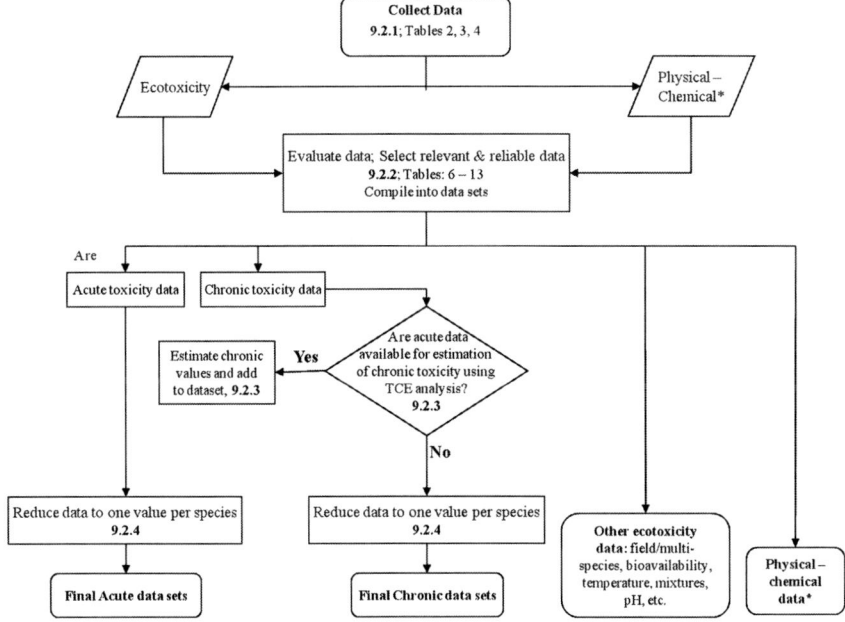

Fig. 3 Data flow chart. For details on each process in the chart, see the tables and sections that are referenced (listed in *bold*)

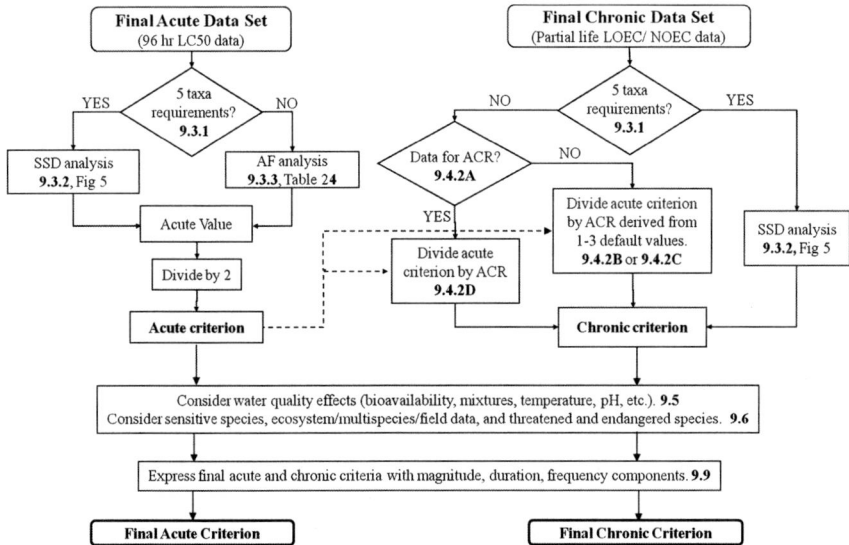

Fig. 4 Criteria derivation flow chart. To begin criteria derivation, see Sections 9.3, 9.4, and Fig. 3. For details on each process in the chart, see the tables and sections that are referenced (listed in *bold*). If plants/algae are most sensitive, refer to Section 9.4.3 instead

9.2.1 Data Collection

The sources listed in Table 3 are recommended for use in collecting the physical–chemical and ecotoxicity data for the pesticide of concern. This is not an exhaustive list, but does contain sufficient resources to locate virtually all available physical–chemical and ecotoxicity data for a given pesticide. Table 4 gives web addresses for electronic resources. Table 2 lists the kinds of physical–chemical and ecotoxicity data that should be collected.

A primary reason for obtaining and using physical–chemical data is to aid in the interpretation of toxicity data studies (Section 9.2.2B). Other such data that is included in Table 2 is useful after criteria are derived, or for other aspects of data interpretation. The LOGKOW database is recommended to serve as the source for K_{OW} values (Sangster Research Laboratories 2004).

Ecotoxicity data are needed and should include aquatic organism studies exposed to pesticides via water. No terrestrial toxicity data (exception in Section 9.2.1B) are used, including laboratory rat and mice studies, or studies with in vitro exposures of organs or tissues (i.e., were not whole-body exposures). Because the UCDM is for derivation of criteria in the United States, only data for freshwater species that are members of families with reproducing populations in North America will be used for criteria derivation, but all data collected may be used for deriving ACRs. USEPA guidelines list species resident in North America (USEPA 1985) in appendices. Literature searches should go back far enough to cover the time period from when the pesticide was first developed to the present.

All original study reports should be obtained from regulators, peer reviewed literature and other sources. Unpublished study reports can be collected from USEPA by reviewing a Re-registration Eligibility Decision (RED) report, and other regulatory documents and requesting appropriate studies through the Freedom of Information Act (FOIA). Local state pesticide regulatory agencies may have copies of such unpublished reports as well (Table 3). Data from regulatory agencies will probably constitute most of the high-quality toxicity studies available, especially for compounds having limited data. Information from regulators should be among the first requested, since is among the most useful, but may require weeks or months to secure.

The rest of this section provides definitions of, and specific guidance on, what kinds of ecotoxicity data should be collected.

A. Single-Species Laboratory Aquatic Toxicity Data

Single-species laboratory aquatic toxicity data are used for criteria calculation in the UCDM. They are derived from laboratory tests with aquatic species and involve aqueous exposures (data from sediment, topical, or oral exposures are not collected or used). Field and multispecies data (including systems with both water and soil/sediment) will only be considered later. Single-species laboratory aquatic toxicity data may be acute or chronic, have several endpoints, and may be expressed in different terms, as described below.

Definitions of Acute and Chronic Toxicity Data

Acute:

(1) Crustacean or insect tests with exposures lasting 24–96 h; (RIVM 2001; Siepmann and Finlayson 2000; USEPA 1985, 2003a),
(2) Fish, mollusk, or amphibian tests with exposures lasting 96 h (RIVM 2001);
(3) Shellfish embryo, larval, or older life-stage tests with exposures lasting 96 h (USEPA 1985, 2003a).

Endpoints measured in plant/algae toxicity tests are usually growth and reproduction, and are generally associated with chronic toxicity. Therefore, explicit definitions for acute plant/algae tests are not included.

Chronic (from USEPA 1985, 2003a):

(1) Plant/algae, single-celled organism tests of any exposure duration;
(2) Any test that takes into account the number of young produced, regardless of exposure duration;
(3) Full life-cycle exposure tests (ranging from 7 days for mysids to 15 months for salmonids);
(4) Partial life-cycle exposure tests (all major life stages exposed for less than 15 months; specifically for fish species that require more than a year to reach sexual maturity);
(5) Early life-stage exposure tests (ranging from 28 to 60 days; also specifically for fish).

Toxicity Values

For derivation of acute criteria, LC_{50} or EC_{50} values are obtained from acute toxicity tests. Chronic criteria or ACRs are used to derive MATCs. Chronic data, expressed as EC_x values (from regression analysis), may be used for criteria derivation only if studies are available to show what level of x is appropriate to represent a no-effect level.

If not reported in a study, LC_{50} or EC_{50} values may be calculated if raw data are available. Similarly, MATC values can be calculated as the geometric mean of the NOEC and LOEC. If NOEC or LOEC values are not stated in a report, but data were evaluated statistically, then the following calculations may be made (based on RIVM 2001):

(a) The highest reported concentration not statistically different from the control ($p < 0.05$) is the NOEC; the NOEC is needed for calculation of the MATC.
(b) The lowest reported concentration that is statistically different from the control ($p < 0.05$) is the LOEC; the LOEC is needed for calculation of the MATC.

(c) For a MATC expressed as a range of values, the NOEC is the lower value, the LOEC is the higher value and the MATC may be calculated as the geometric mean, as described previously.

Toxicity Endpoints

Appropriate endpoints for criteria derivation are those that measure survival, growth, or reproductive effects. This includes measures of immobility, as well as population-level endpoints, such as r (intrinsic rate of population growth) and λ (factor by which a population increases in a given time). Endpoints other than survival, immobility, growth, reproduction, r, or λ may be used in criteria derivation if those endpoints have been linked to effects on survival, growth, or reproduction. For example, if it is determined from a study that an 80% effect on AChE inhibition is significant (in either an acute or a chronic exposure), and if 80% AChE inhibition is shown to lead to mortality *for that species*, then an IC_{80} value (concentration that causes 80% inhibition compared to the control) may be used as a toxicity value in criteria derivation. Alternatively, if that same study disclosed a LOEC that represents 80% reduction from control, then the corresponding MATC value from that study may be used in criteria derivation, or for derivation of an ACR. It is important to emphasize that levels of sub-lethal effects that lead to reductions in survival, growth, or reproduction are species specific. If no data are available to link effects, such as endocrine disruption, enzyme induction, enzyme inhibition, behavioral effects, histological effects, stress protein induction, changes in RNA or DNA levels, mutagenicity, and carcinogenicity to survival, growth, or reproduction, these data are not to be used directly for criteria derivation.

B. Other Ecotoxicity Data

Single-species laboratory aquatic toxicity data (described in Section 9.2.1A) will be used directly for criteria derivation. Other data described in the next three sections (9.5, 9.6, and 9.7) may be used to check or modify criteria, depending on availability.

Multispecies (Field/Semi-field/Laboratory) Data

Multispecies data are not used directly for criteria derivation. However, they should be collected because multispecies laboratory, field, or semi-field data are used in Section 9.6.2 for comparison to criteria derived from single-species data (OECD 1995a; RIVM 2001) and may provide justification for adjustment of a final criterion (RIVM 2001; USEPA 1985, 2003a; Zabel and Cole 1999).

Water Quality Effects Data

After criteria are derived from single-species studies, water quality effects will be considered; these effects include the following: the effects of suspended particulate matter on bioavailability; the effects of pesticide mixtures; and the effects of temperature, pH, or other water quality parameters on toxicity. Water quality effects data should be collected and it is recommended that Section 9.5 be reviewed first to know what kind of studies will be useful.

Terrestrial and Human Health Data

Although these criteria are not intended for protection of human or terrestrial life, a separate section is included to address bioaccumulation or secondary poisoning in terrestrial organisms that may be indirectly exposed from feeding on aqueous species that have pesticide in their tissues. This section is only required if the compound is likely to bioaccumulate; therefore, this section should be reviewed before collecting the required wildlife and human health data (Section 9.7.1).

9.2.2 Data Evaluation

In this section, instruction is given for how to determine if data are relevant and reliable for use in deriving water quality criteria.

A. Physical–Chemical Data

Evaluate physical–chemical data according to whether it was obtained by an appropriate method that was properly used. In Table 6, we present acceptable methods for determination of a number of physical–chemical parameters other than K_{OW}. Table 7 shows acceptable methods for determination of K_{OW} values. The methods shown in Table 7 are listed in order of preference; computational methods should only be used if no measured data are available. The recommended values in the LOGKOW database (Sangster Research Laboratories 2004) may be used without further review because they have been thoroughly reviewed before inclusion in the database. Physical–chemical parameters reported by manufacturers may also be used without further review, because they are widely accepted, and original studies are usually not published. Physical–chemical parameters, determined by methods not shown in Tables 6 and 7 (or equivalent methods), should be used with caution.

B. Ecotoxicity Data

The physical–chemical data are used to evaluate ecotoxicity studies. Water solubility data are needed to compare to test concentrations to ensure that none of the compound precipitated. Half-life ($t_{1/2}$), partition coefficients (K_{OC}, K_{OW}, K_H), and vapor pressure data are important to determine if a tested compound will dissipate rapidly in a static test, making a flow-though exposure more appropriate. Tests that contain toxicity values greater than $2x$ the geometric mean of available water solubility values for the pesticide are not useful even as supporting information, and can be eliminated without further consideration. For compounds with log K_{OW} values between 5 and 7, laboratory tests should use feeding regimes that minimize or eliminate interaction of pesticide with food particles.

Ecotoxicity data are evaluated for relevance and reliability. For the single-species tests, a relevance score is calculated using the rating system presented in Table 8, and a rating of *R* (relevant), *L* (less relevant), or *N* (not relevant) is assigned, based on the scale shown in Table 13. Tests that score <70 (i.e., rating = *N*) do not need to be further evaluated, although it is useful to create a brief record of the citation and list of the relevance parameters not fulfilled. All single-species tests with a relevance score ≥ 70 (i.e., rating = *R* or *L*) should be summarized. A data summary sheet, such as the one shown in Table 5, helps to ensure that all relevant information is drawn from each study. In the data summary sheets, it is important to report all toxicity values from different time points, endpoints, or repeated tests. The most appropriate values will be selected later in the data reduction procedures. Using the data in these summary sheets, and the rating systems shown in Tables 9 and 10, one can evaluate single-species aquatic ecotoxicity studies on two aspects of reliability: (1) documentation and (2) acceptability. The mean of the documentation and acceptability scores is used to calculate the reliability score of a study, and the rating system in Table 13 can be used to assign a score of *R* (reliable), *L* (less reliable), or *N* (not reliable).

The next step is to evaluate other types of aquatic toxicity tests (i.e., multispecies laboratory/field, microcosm, mesocosm) on documentation and acceptability, using parameters in Table 11. Terrestrial toxicity studies are evaluated solely based on the documentation parameters that appear in Table 12. Reliability ratings are assigned to each study of *R*, *L*, or *N*, based on the scale presented in Table 13. Specific instructions for rating various kinds of ecotoxicity studies are given immediately below.

Single-Species Laboratory Studies (Aquatic Species with Aqueous Exposures)

(1) Rate relevance using the scoring system in Table 8; if the relevance score is ≥ 70, proceed to the following steps; if the relevance score is <70, the test is not usable and does not need to be further evaluated.
(2) Fill in data summary sheet (Table 5).
(3) Rate documentation using the scoring system in Table 9.
(4) Rate acceptability using the scoring system in Table 10.
(5) Average the scores from steps 3 and 4 for an overall reliability rating.
(6) Assign the study to a category based on relevance and reliability scores according to the contents of Table 13.
(7) Use studies rated RR for criteria derivation; use studies rated RL, LR or LL as supporting data; do not use studies receiving *N* ratings.

Aquatic Outdoor Field Data/Indoor Model Ecosystems (Including Microcosms/Mesocosms), Multispecies Data

(1) Rate documentation and acceptability using the scoring system in Table 11.
(2) Assign a reliability rating of *R*, *L*, or *N* using the scoring system in Table 13.
(3) Use studies rated *R* or *L* to evaluate potential ecosystem effects (Section 9.6.2); do not use studies rated *N*.

Terrestrial Wildlife Data

(1) Rate documentation using the scoring system presented in Table 12.
(2) Assign a reliability rating of *R*, *L*, or *N*, using the scoring system presented in Table 13.
(3) Use studies rated *R* or *L* to assess potential hazards due to pesticide bioaccumulation (Section 9.7.1); do not use studies rated *N*.

Organize single-species data into tables with at least the genus, species, value(s), and reference. Create different tables for acute data-rated RR, chronic data-rated RR, the supplemental data (rated RL, LL, LR), and data excluded from calculations as part of the reduction process. If a study has results from multiple tests with the same species, report each value as an individual test by the same author. These toxicity values will be combined when data is reduced.

9.2.3 Filling Chronic Toxicity Data Gaps with Estimation Techniques

Chronic data sets may be supplemented by use of extrapolation techniques, called TCE analysis, that produce estimates of chronic toxicity based on acute toxicity data. These data may be used in SSD criteria derivation procedures (Section 9.3.2), but not in an ACR (Section 9.4.2). Values from TCE analysis may be used to fulfill missing taxa requirements for generation of a chronic SSD. TCE analysis requires acute mortality data that has three components: exposure concentration, degree of response, and time course of effect. Procuring the values for these components requires having access to raw toxicity data that includes exposure concentrations and measurements of mortality at multiple time points.

If there are appropriate acute data that can be used to estimate chronic data for use in the SSD, perform the TCE analysis by using USEPA's acute-to-chronic estimation software (ACE, v. 2.0, Ellersieck et al. 2003, which is available for free download at http://www.epa.gov/ceampubl/fchain/index.htm). This software comes with a user's manual that fully explains the models used, explains how to choose a model, describes model limitations, and gives guidance on how to use the software. The ACE program output provides estimated toxicity values for a range of mortality levels and a range of chronic exposure periods. For the accelerated life testing model, a 1% mortality level is recommended to represent a NOEC, whereas for the multifactor probit analysis and linear regression analysis models, a 0.01% effect level is recommended (Ellersieck et al. 2003). The exposure period should be selected to reflect a full life-cycle of the organism used in the acute study.

9.2.4 Data Reduction

For criteria derivation, data must be reduced such that each species has one representative data point in the final data set. When there is more than one toxicity value for a species, data are reduced to a single SMAV or SMCV.

Following are the specific data reduction procedures:

(1) Calculate the SMAVs/SMCVs as the geometric mean of toxicity values from one or more acceptable tests with the same endpoints (ANZECC and ARMCANZ 2000; ECB 2003; OECD 1995a; RIVM 2001; USEPA 1985, 2003a).
(2) If data are available for life stages that are at least a factor of two more resistant than another life stage for the same species, then use the data for the more sensitive life stage to calculate the SMAV because the goal is to protect all life stages (RIVM 2001; USEPA 1985, 2003a).
(3) If data are available for multiple appropriate endpoints (see Section 9.2.1A) for one species, then use the data for the most sensitive endpoint (ANZECC and ARMCANZ 2000; ECB 2003; OECD 1995a; RIVM 2001).
(4) If a NOEC is not explicitly reported in chronic toxicity studies, but statistical analysis was done, the NOEC may be determined as the highest reported concentration not statistically different from the control ($p < 0.05$; RIVM 2001), the NOEC is not used in criteria derivation, but is needed for calculation of the MATC.
(5) Similarly, if a LOEC is not explicitly reported in chronic toxicity studies, it may be determined as the lowest reported concentration that is statistically different from the control ($p < 0.05$); the LOEC is not used in criteria derivation, but is needed for calculation of the MATC.
(6) If a MATC is not reported, it may be calculated as the geometric mean of the NOEC and LOEC.
(7) If no toxicity values were reported, but raw data are available, calculate toxicity values using appropriate statistical methods (ECB 2003).
(8) If a MATC is expressed as a range of values, recalculate the MATC as the geometric mean of the high and low values (RIVM 2001).
(9) If reasons for differences between tests for the same species/endpoints are found, then data may be grouped according to appropriate factors (e.g., pH or temperature; ECB 2003). Selection of the appropriate value to use in criteria derivation should be based on standard test parameters. Tests conducted under non-standard conditions (vs. standard conditions as defined in standard test methods) may be used to derive quantitative relationships between those conditions and toxicity (as in USEPA 1985, 2003a). If such a relationship is established, then toxicity values derived under non-standard conditions may be translated to standard conditions and added to the criteria derivation data set. If no quantitative relationship can be derived, then tests conducted under non-standard conditions should not be used for criteria derivation, but may be used as supporting information.
(10) If data are available for multiple time points from crustacean or insect acute toxicity studies, use the latest time point (i.e., 96-h tests are preferred over tests of <96 h).
(11) For a given species, use data from flow-through tests in which concentrations were measured if such data are available. If such data are not available, then data from static or static-renewal tests and/or tests in which concentrations

were not measured may be used as long as they are rated otherwise relevant and reliable.

9.2.5 Data Graphing

Graphing the toxicity data is useful for evaluation of multi-modality and outliers. First, a histogram of the frequency distribution should be constructed (see Section 3.1.1 for examples). The distribution should be examined for multi-modality (see Section 9.3.2E, part a) or for outliers. Toxicity values should be double-checked for errors, especially if any toxicity values appear to be outliers. A multi-modal distribution may be more easily seen when graphing a cumulative frequency distribution. This can be done as part of the SSD fitting in the next sections, or a graph of cumulative frequency vs. log concentration can be constructed using Equation (29) below. If a distribution is used to calculate a final criterion, a graph of the distribution plotted with the actual toxicity values should be included in the final report.

$$\text{Cumulative frequency} = \frac{\text{rank} - 0.5}{n} \qquad (29)$$

where

rank = position in a set of ordered data (ranked from lowest to highest);
n = sample number.

Once data are collected, evaluated, selected, and reduced, criteria derivation may begin.

9.3 Acute Criterion Derivation

If five acute data requirements can be fulfilled (see below), an SSD will be used to derive the acute criterion in Section 9.3.2. Otherwise an AF will be used in Section 9.3.3.

9.3.1 Data Requirements Applicable to Use of the SSD

Data must first be collected, evaluated, and reduced as described in Sections 9.2, 9.3, and 9.2.4. For derivation of acute or chronic criteria by the SSD procedure, a minimum of five data values from five different test-organism families are required. The procedure specified in Section 9.4.3 should be used to derive chronic criteria for herbicides; however, an acute criterion should be derived using animal data, if possible, for an herbicide. To use the SSD procedure, a data set must include examples of

(a) the family Salmonidae;
(b) a warm water fish;

(c) a planktonic crustacean, of which one must be in the genus *Ceriodaphnia, Daphnia,* or *Simocephalus* from the family Daphniidae;
(d) a benthic crustacean;
(e) an insect (aquatic exposure).

If these five requirements are met, then the SSD procedure, described in Section 9.3.2, may be used to derive the acute criterion. If such data are not available, then the AF procedure, described in Section 9.3.3, may be used.

9.3.2 Derivation of a Criterion Using an SSD

Depending on the number of species mean toxicity values, either the Burr Type III distribution (Section 9.3.2A) or the log-logistic distribution (Section 9.3.2B) will be used. Data from all taxa should be combined for this procedure, but data on plants and algae should be kept separate. From the fitted distribution, it is possible to calculate concentrations that will protect 95% of species with 50% confidence (95:50), 95% of species with 95% confidence (95:95), 99% of species with 50% confidence (99:50), and 99% of species with 95% confidence (99:95). The number that is most robust of these is the one selected to protect 95% of species with 50% confidence. This median 5th percentile estimate is recommended for derivation of the acute criterion. The other estimates may be used if more conservative values are desired, but since they come from the extreme tails of the SSD, they are less reliable.

A. Use of the Burr Type III SSD: For More than Eight Toxicity Values

The criteria can be derived using the SSD procedure described in ANZECC and ARMCANZ (2000). Using any statistical package that is appropriate, fit a Burr Type III distribution (Burr III, reciprocal Weibull, or reciprocal Pareto; Burr 1942) to the data and calculate the 1st and 5th percentile values using the following equations (record to three significant figures):

$$\text{PC}(q) = \frac{b}{\left[\left(\frac{1}{1-q}\right)^{\frac{1}{k}} - 1\right]^{\frac{1}{c}}} \tag{30}$$

where

PC(q) is the protecting concentration that will protect q% of species; thus, the 5th percentile is calculated by setting $q = 95$;
$q =$ percent of species to protect;
b, c, k are fit parameters.

For reciprocal Weibull (for cases when $k \to \infty$)

$$PC(q) = (-\alpha/\ln(1-q))^{1/b} \qquad (31)$$

where

PC(q) and q are as described for Burr III;
α and β are fit parameters.

For reciprocal Pareto (for cases when $c \to \infty$)

$$PC(q) = x_0(1-q)^{1/\theta} \qquad (32)$$

where

PC(q) and q are as described for Burr III;
x_0 and θ are fit parameters.

It is acceptable to use any statistical package that can fit Burr Type III distributions to accomplish this calculation and the calculation of confidence limits discussed in Section 9.3.2C. The BurrliOZ program, which was developed specifically for use in deriving target values (criteria) in the ANZECC and ARMCANZ (2000) methodology, is available for free from the CSIRO website (http://www.cmis.csiro.au/Envir/burrlioz/). The BurrliOZ software comes with the caution that for data sets of eight or fewer toxicity values, there will be great uncertainty in the calculated values. The software authors provide a procedure to follow in such cases, which has been modified for the UCDM and is presented in Section 9.3.2B.

To complete the Burr Type III SSD analysis, perform the fit test as in Section 9.3.2D and calculate confidence limits as described in Section 9.3.2C.

B. Use of the Log-Logistic SSD: For Eight or Fewer Toxicity Values

When there are eight or fewer toxicity values in the data set, preference should be given to using the log-logistic distribution, rather than the Burr Type III distribution. Note: the BurrliOZ software contains a specific procedure that compares the fits of the log-logistic distribution and the Burr Type III distribution, and then uses the distribution that appears to give the better fit. The procedure described here is a modification of that procedure and is the one recommended for use in the UCDM.

Using any capable statistics package, fit the data to a log-logistic distribution. An example of such a program is ETX v.1.3 (Aldenberg 1993) for which software can be obtained from RIVM by contacting info@rivm.nl. Once the fit parameters

(α and β) have been determined, the following equation can be used to determine 1st and 5th percentile values:

$$p = \frac{100}{1 + \exp(-[\ln(x) - \alpha]/\beta)} \tag{33}$$

where

p = percentage of species unaffected at x; set $p = 1$ to calculate the 1st percentile; $p = 5$ for the 5th percentile;
x = toxicity value at p;
α = sample mean (of $\ln(x)$);
$\beta = k_L \cdot s_n/C_5$.

and

k_L = extrapolation constant; dependent on sample size; selected for either median or lower 95th percentile estimate (see Table 23);
s_n = sample standard deviation (of $\ln(x)$); n = sample size;
C_5 = constant = 2.9444.

Table 23 Extrapolation constants, k, for median and lower 95% confidence limit estimates of the 5th percentile value using a log-logistic distribution (taken from Aldenberg and Slob 1993)

n	Median	Lower 95% confidence limit
2	2.49	27.7
3	2.05	8.14
4	1.92	5.49
5	1.85	4.47
6	1.81	3.93
7	1.78	3.59
8	1.76	3.37
9	1.75	3.19
10	1.73	3.06
11	1.72	2.96
12	1.72	2.87
13	1.71	2.80
14	1.70	2.74
15	1.70	2.68
20	1.68	2.49
30	1.66	2.28
50	1.65	2.10
100	1.64	1.95
200	1.63	1.85
500	1.63	1.76
∞	1.62	1.62

Note: some software uses log(x) in place of ln(x) in Equation (33), and to calculate α and β, such as the ETX v. 1.3 software. If using α and β calculated from log(x), be sure also to use log(x) in the Equation (33) instead of ln(x).

To complete the log-logistic SSD analysis, calculate confidence limits as described in Section 9.3.2C and perform the fit test as described in Section 9.3.2D.

If the fit of the data to the log-logistic distribution passed the fit test ($p > 0.05$), then this distribution should be used to calculate the 1st and 5th percentile values for the data set. If the log-logistic distribution fails the fit test, then use the procedure presented in Section 9.3.2A to fit the data to the Burr Type III distribution.

C. Calculation of Confidence Limits

The values calculated in Section 9.3.2A represent median estimates of the 1st and 5th percentiles. To estimate the lower 95% confidence limit for these estimates, utilize the following bootstrapping technique (CSIRO 2001):

(1) Resample the original data set, with replacement, to create a new data set of the same size as the original set, and calculate 1st and 5th percentile values from the new data set. Repeat this resampling and recalculation procedure 200–1000 times. At least 501 resamplings are recommended (ANZECC and ARMCANZ 2000), fewer will give a less certain estimate; more will give a more certain estimate, but requires more calculation time.
(2) Order the bootstrapped estimates from lowest to highest (separately for the 1st and 5th percentile SSD estimates) and select the 5th percentile value; this represents the lower 95% confidence limit estimate of the 1st or 5th percentile of the SSD.

These procedures can be accomplished using the program BurrliOZ v. 1.0.13 (CSIRO 2001). The software can be obtained at http://www.cmis.csiro.au/Envir/burrlioz/. Also the ETX v.1.3 software (Aldenberg 1993) can be used to calculate the 95% confidence limit for the 5th percentile estimate for the log-logistic distribution, but not for 1st percentile. The latter estimate may be omitted for the log-logistic distribution since the other three are likely to be more useful because they have less uncertainty.

D. Checking the Goodness of Fit of the SSD

The following procedure is used to evaluate the goodness of fit of the SSD to the toxicity data. If the BurrliOZ software is used to fit an SSD, it provides an evaluation of the goodness of fit based on a statistical method called maximum likelihood estimation. A different fit test is used in the UCDM that is based on cross-validation. In general, the cross-validation approach starts by omitting the first data point and then refitting the distribution. Then the probability of the omitted point is estimated using the new distribution. This is done for each data point in turn, and the combined results for all points in the data set are examined for a significant lack of fit, using Fisher's combined test (outlined below).

The distribution will have been fitted using a sample of n species toxicity values, expressed as concentrations, and these can also be called x values (as in plotting y vs. x). To begin, the distribution is refitted using the data set that *omits* the point x_i, and the subsequent distribution function is called F_{-i}. The next step is to assess the placement of the omitted point within this distribution, which is called $F_{-i}(x_i)$. Solving for $F_{-i}(x_i)$ provides the corresponding probability for x_i (which would also be called the y value). After the distribution is refitted, the results window in the BurrliOZ software allows entry of a concentration (x_i) and then provides the corresponding percentile, to solve for $F_{-i}(x_i)$. $F_{-i}(x_i)$ can then be calculated for each data point.

Then let

$$p_i = 2 * \min(F_{-i}(x_i), 1 - F_{-i}(x_i)) \tag{34}$$

where "min" indicates the use of the minimum of either $F_{-i}(x_i)$ or $1 - F_{-i}(x_i)$.

Next, Fisher's combined test can be applied to calculate a Chi squared statistic of the form

$$X^2_{2n} \sim -2 \sum i \ln(p_i) \tag{35}$$

If any one of the data points is insufficiently fitted, then the test is capable of rejecting the hypothesis that the data come from the fitted (Burr Type III) distribution. Once all of the p_i values have been calculated, the Chi squared statistic (X^2) can be calculated. (In Excel the significance of the Chi squared statistic is calculated with the command of the same name:

CHIDIST, with the fields $(x,$ deg freedom),

where $x = -2 \sum_i \ln(p_i)$ and deg freedom is the degrees of freedom or n, the number of p_i values.)

The closer the resulting value for X^2_{2n} is to 1, the better the fit. When the result for $X^2_{2n} <0.05$, there is a significant effect from the substitution and a 95% probability of a significant lack of fit. If there is a significant lack of fit, the data should then be critically examined and checked for multi-modality. If multi-modality is evident, a different procedure may be used as described in Section 9.3.2E and Fig. 5.

The University of California-Davis Methodology for Deriving Aquatic Criteria 111

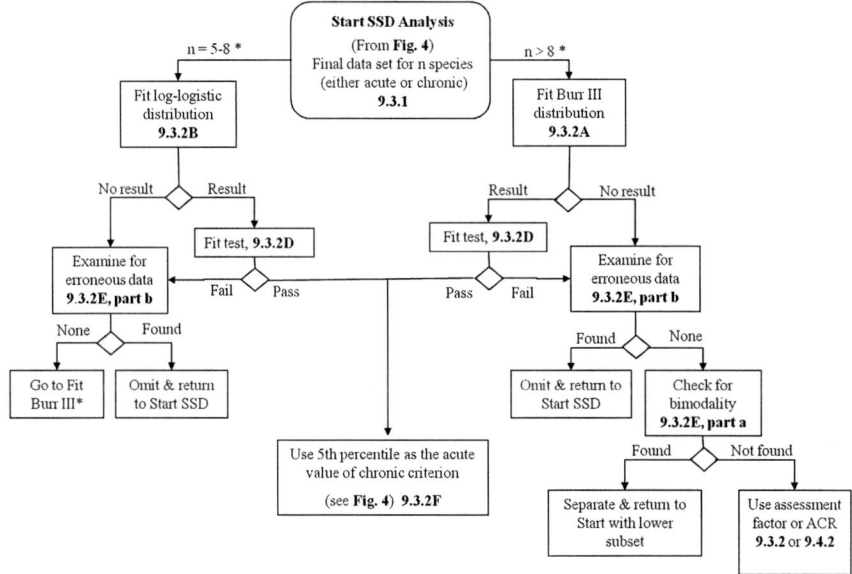

Fig. 5 SSD flow chart. The five taxa requirements should be met before a distribution is used (except for a subset of a multi-modal data set, Section 9.3.2E, part a). For details on each process in the chart, see the referenced tables and section (listed in *bold*)

E. Procedure if the SSD Fails the Fit Test

If the full data set cannot be fitted to an SSD (procedure described in Sections 9.3.2A and 9.3.2B), examine the data for multi-modality and/or outliers as outlined below. If appropriate, reanalyze using the appropriate procedure for the remaining number of data points and the fulfilled taxa requirements (Fig. 5).

(a) Examine data for multi-modality. If an SSD cannot be fitted and visual inspection indicates that the SSD is multi-modal, assess whether the subsets are divided by a justifiable parameter (such as by taxa). If justifiable, divide the data into subsets and use the subset containing the lowest toxicity values (ANZECC and ARMCANZ 2000) to calculate the criterion. This is easily done in conjunction with the data plotting step explained in Section 9.2.5. A distribution can be fitted to a subset that does not contain the five taxa requirements, provided that the original data set fulfilled these requirements and the final subset contains at least five data points.
(b) Double-check the toxicity values to ensure that they do not contain mistakes (i.e., typographical or transcriptional errors), and review the original studies to ensure that all test conditions were appropriate. The need to remove outliers is considerably reduced when using the Burr Type III distribution with the

BurrliOZ software (CSIRO Biometrics; Campbell et al. 2000). If a satisfactory fit cannot be obtained with the full data set, critical examination of data is essential to determine if any one point is an outlier. An individual point that causes the SSD to not fit probably represents an erroneous data point (e.g., above the water solubility of the compound or below the analytical detection limit). If errors in data are found, remove them from the data set and use the data that remains. Removal of data from the SSD may also be justified if there is a scientific rationale that explains why the outlier(s) do(es) not belong in the same SSD as the mainstream data (similar to discriminating based on multi-modality, e.g., a resistant strain of mosquitoes). This approach is reasonable because, as with all criteria derived from the UCDM, criteria will be evaluated to determine if they will provide adequate protection (Section 9.6).

(c) If removal of data is not justifiable, and it is not possible to fit an SSD, one should use the AF procedure (Section 9.3.3). When only 5–8 data points are available, the AF procedure should be used instead of an SSD, because the data are most likely to be multi-modal, but there are not enough data to clearly separate the lower subset. However, with eight or fewer data points, users should attempt to fit the data to a log-logistic distribution before the Burr Type III distribution is used (see Section 9.3.2B).

F. Calculation of a Criterion from the 5th Percentile Value

For the acute criterion

$$\text{The recommended criterion} = (\text{5th percentile value at 50\% confidence level}) \div 2$$

For the chronic criterion

$$\text{The recommended criterion} = \text{5th percentile value at 50\% confidence level}$$

Alternatively, more conservative criteria may be derived from other percentile or confidence levels.

The number of significant digits in the final criterion should be consistent with known variability in the calculated criteria. There are three guidelines that may be used for choosing the number of significant digits to report in the final criteria. The first guideline is that the calculated criteria should not be expressed with more significant figures than those reflected in the original toxicity data; this guideline can be used for any data set, regardless of which criteria calculation procedure was used. The second guideline is to compare the median estimate to the lower 95% confidence limit; this guideline can only be used if a distribution could be fit to the data set. If one uses the median estimate as the criterion, then the 95% confidence limit can be used as a guide to determine the appropriate number of significant figures.

The first digit in the median estimate that differs from the lower 95% confidence limit denotes the last significant digit that should be reported. For example, if the median estimate is 0.143 µg/L, and the lower 95% confidence limit is 0.0752 µg/L, then one significant figure would be reported for the median estimate (i.e., 0.1 µg/L) because the first significant digit of the median estimate is different from that of the lower 95% confidence limit. The third guideline can also only be used if a distribution was fit to the data set. The third guideline is to use the 5th percentile values generated from omitting data sets during the fit test (Section 9.3.2D) to estimate the uncertainty of the calculated criteria; the last digit that is relatively variable among these estimates can be used to determine the last significant digit.

If toxicity is quantitatively related to a water quality parameter, follow the procedures given in Section 9.5.3 for appropriate expression of the criterion. The criteria should be checked against the individual toxicity values in the data sets (Section 9.6.1) to increase confidence that all represented species will be protected.

9.3.3 Derivation of an Acute Criterion Using an AF

If data requirements for the SSD procedure cannot be met, or an SSD cannot be fit, then the AF procedure must be used to derive criteria. To do this, one must divide the lowest SMAV from the data set by a factor (Table 24). The magnitude of the factor is dependent on the number of data requirements met, and at least one of the available, acceptable data values must be on a test species from the family Daphniidae in the genus *Daphnia*, *Ceriodaphnia*, or *Simocephalus,* or a criterion cannot be calculated. Each of the additional data values must fulfill a different taxonomic requirement in the SSD minimum data set (Section 9.3.1). When properly used, each additional value builds toward completion of the minimum SSD data set. The resulting value represents an estimate of the median 5th percentile value of the SSD.

Table 24 AFs to apply to lowest acute toxicity values in data sets that meet fewer than five of the taxa requirements

Number of taxa requirements	AF
1	57×10[a]
2	36
3	7.8
4	5.1
5[b]	3.8

[a]The factor 57 was derived from pesticide data; the 10 is an additional factor assessed to protect against cases in which Daphnids are among the most tolerant species.
[b]This factor is provided for use if the data requirements are met, but the SSD cannot be fit.

$$\text{Acute criterion} = (\text{lowest value in data set} \div \text{AF}) \div 2$$
$$= \text{estimated 5th percentile value} \div 2$$

If toxicity is quantitatively related to a water quality parameter, the procedures in Section 9.5.3 should be followed for appropriate expression of the criterion.

It should be noted that these AFs were formulated with data from organic insecticides. Because some molluscicides, miticides, and fungicides have similar properties, these factors would serve as a reasonable means of estimating criteria for them, as well. These factors should not be used with metal-based pesticides. For herbicides (or plant-toxic agents), however, another procedure should be used as described in Section 9.4.3. The AFs in Table 24 may be updated and recalculated as more criteria are generated. Data sets that meet the five taxa requirements for use of an SSD may be added to those originally used in Section 3.2 to calculate updated AFs.

9.4 Chronic Criterion Derivation

If the required five chronic data values can be fulfilled, the SSD approach should be used to derive the chronic criterion (described below); otherwise, an ACR may be used (Section 9.4.2).

9.4.1 Derivation of a Chronic Criterion Using an SSD

If at least five chronic toxicity data values are available for species from five different families, as described in Section 9.3.1 (either from direct measurements or from TCE estimates as described in Section 9.2.3), then the instructions in Section 9.3.2 may be followed to determine chronic 5th and 1st percentile values at various confidence levels. If such data are not available, then proceed to Section 9.4.2 for a description on how to derive chronic criterion by applying an ACR to the acute criterion. For herbicides, the chronic value is derived using the procedure outlined in Section 9.4.3. A chronic value derived using the SSD procedure does not require any further adjustment by a safety factor because this value is derived from long-term no-effect toxicity values.

9.4.2 Derivation of a Chronic Criterion Using an ACR

When chronic data are lacking, ACRs can be used to extrapolate from acute to chronic toxicity. Preferably, ACRs from measured (experimental) toxicity values should be used if they are available. ACRs can be derived by following the procedures explained in Sections 9.4.2A, 9.4.2B, and 9.4.2C (based on ANZECC and ARMCANZ 2000; USEPA 1985, 2003a). If sufficient data are not available to calculate ACRs from measured toxicity data, a default value can be used and is provided in Section 9.4.2C. Resulting criteria should be checked against the individual toxicity values in the chronic toxicity data set to ensure protection of all represented species (Section 9.6.1).

A. Use of Single-Chemical, Multispecies ACR Based on Measured Toxicity Values

This procedure requires acute and chronic data from organisms in at least three different families, including a fish, an invertebrate, and at least one other acutely sensitive species. For each acceptable chronic value (MATC) that has at least one corresponding appropriate acceptable acute value, an ACR can be calculated by dividing the flow-through acute test value by the chronic value. Static tests are acceptable for midges, daphnids, and other zooplankton. For fish, the acute test(s) should be conducted with juvenile or younger fish. For all species, the acute test(s) should be part of the same study and use the same dilution water as used in the chronic test. If there are multiple acute tests that are equally appropriate, then the geometric mean of the toxicity values should be used. If acute tests were not conducted as part of the same study, but were conducted as part of a different study in the same laboratory and dilution water, then the values from them may be used. If no such acute tests are available, results of acute tests conducted in the same dilution water in a different laboratory may be used. If there are not enough freshwater data to fulfill the ACR data requirements, then saltwater species may be used because freshwater and saltwater ACRs have been shown to be comparable (USEPA 1985), and this approach has been accepted in numerous criteria derivation methods (Siepmann and Finlayson 2000; USEPA 1980a, b, c, d, 2003d, 2005a).

It is possible that there may be acceptable ACR data from several studies for an individual species. If this is the case, the ACRs should first be calculated separately for each study, resulting in several ACRs for a given species. Second, the SMACR can be determined by calculating the geometric mean of all ACRs available for an individual species. For some materials, the SMACRs seem to be the same for all species, but for others the ratios seem to increase or decrease as the SMAV increases. Thus, the final multispecies ACR can be obtained in one of three ways, depending on the data available:

(1) If the SMACRs seem to increase or decrease as the SMAVs increase, then calculate the final multispecies ACR as the geometric mean of the SMACRs for species whose SMAVs are close to the acute criterion (this includes species whose SMACRs are within a factor of 10 of the SMACR of the species whose SMAV is nearest the acute median 5th percentile estimate).
(2) If no major trend is apparent and the SMACRs for all species are within a factor of 10, then calculate the final multispecies ACR as the geometric mean of all of the SMACRs.
(3) If the most appropriate SMACRs are less than 2.0, and especially if they are less than 1.0, then acclimation has probably occurred during the chronic test(s). In this situation, assume the final ACR to be 2.0, so that the chronic criterion is equal to the acute criterion.

If the data requirements of this section cannot be met, or if the ACR cannot be obtained by one of these methods (1, 2, or 3 above), then derive the final multispecies ACR by the procedure given below, in Section 9.4.2B.

B. Use of Single-Chemical, Multispecies ACR Based on Measured Toxicity Values and/or Default ACR Values

If not enough data are available for calculation of an ACR according to the procedure in Section 9.4.2A, then default ACR values can be added to the ACR data set to give a third ACR for use in producing a geometric mean (USEPA 2003a). The default ACR is 12.4 and it is described in the next Section (9.4.2C). If there are two measured ACRs, then calculate the geometric mean of the two measured ACRs and one default ACR to derive the final ACR. If there is one measured ACR, then calculate the geometric mean of the measured ACR and two default ACRs to derive the final ACR. If no measured ACRs are available, then three default ACRs are used, of which the geometric mean would be the default ACR itself of 12.4.

C. Use of a Default ACR

A default ACR of 12.4 was calculated specifically for use with pesticides by using the procedure described in Section 3.2.5. This default ACR was calculated using experimental ACRs from the literature, all of which were from studies performed on organic insecticides. Some molluscicides, miticides, and fungicides have properties similar to organic insecticides, and this default ACR would serve as a reasonable means of estimating criteria for such compounds. The default ACR should not be used with metals (for possible alternatives see the discussion on the derivation of the default ACR in Section 3.2.5 and Host et al. 1995). However, for herbicides (or if plants are most sensitive to the pesticide in question), the procedure described in Section 9.4.3 should be used instead. The default ACR may be revised if (1) the data sets collected according to the UCDM lead to different ACR values; (2) previously calculated ACRs are shown to be invalid based on data sets collected according to the UCDM; or (3) additional pesticide ACR values become available in other USEPA criteria documents (or similar thoroughly vetted criteria documents). In any of these events, the default ACR should be recalculated as the 80th percentile value of the new set of experimental ACRs. Table 21 shows the current set of ACRs that are used to calculate the default value. Any future revisions of the value should start with this data set.

D. Chronic Criterion Calculation

To calculate the chronic criterion, the acute 5th (or 1st) percentile value (derived by the SSD procedure or estimated by the AF procedure) is divided by the final multispecies ACR (derived by one of the three methods in Sections 9.4.2A, 9.4.2B, and 9.4.2C). This approach is equivalent to that used in the USEPA methodologies, in which the FAV (i.e., the 5th percentile value) is divided by the ACR, to derive the chronic criterion (USEPA 1985, 2003a):

$$\text{Chronic Criterion} = (\text{Selected percentile value}) \div \text{ACR}$$

If toxicity is quantitatively related to a water quality parameter, then follow procedures in Section 9.5.3 for appropriate expression of the criterion.

9.4.3 Derivation of a Chronic Criterion for a Herbicide

When the pesticide of concern is a herbicide, alga or vascular aquatic plant data must be included. Since life cycles of plants vary widely, and procedures for conducting toxicity tests with plants are not well developed, explicit definitions for acute plant tests are not included in the UCDM. Therefore, plant data are only used to derive the chronic criterion. The procedure to derive a chronic criterion for herbicides is as follows.

If the chemical is a herbicide, or plants are the organisms that are most sensitive to it:

(1) Fit an SSD with only alga or vascular aquatic plant data if there are data from at least five different species that were rated RR.
(2) If there is not enough data to do the SSD as described above, then use the lowest NOEC value from an important alga or vascular aquatic plant species that has measured concentrations and the endpoint is biologically relevant.

Few criteria have been derived for herbicides and, as a result, approaches are not as well described for this group as are criteria calculation procedures for other pesticides. This is an area in which new approaches are currently being developed. The Minnesota Pollution Control Agency (Angela Preimesberger) is working on criteria development for herbicides. Mark Hanson of University of Manitoba is working on guidance for interpreting plant/algal toxicity data. They and other agencies may be good resources to consult about how to best work with individual plant data sets.

9.5 Incorporation of Water Quality Effects into Criteria Compliance

If the toxicity of a chemical can be quantitatively related to one or more water quality characteristics, then the criteria can either be expressed in the form of equations that quantify the relationships, or the relationship can be used to determine site-specific compliance with criteria. For organic pesticides, the water quality characteristics of primary concern are as follows: the effects of suspended particulate matter on bioavailability, the effects of pesticide mixtures, and the effects of temperature, pH, or other parameters on toxicity. Bioavailability is addressed in Section 9.5.1; methods for compliance determination, in cases where pesticide mixtures exist, is presented in Section 9.5.2; and methods also used by USEPA (1985, 2003a) for expression of criteria in the form of equations relating pH, temperature, or other parameters to toxicity are presented in Section 9.5.3.

9.5.1 Bioavailability

If significant levels of suspended and/or dissolved solids co-occur with pesticides in a water body, then it may be desirable to consider the effects of solids on the bioavailability of pesticides in determining compliance with derived criteria. The following approach is recommended:

(1) In the water column, pesticides may be sorbed to solids or dissolved solids, or freely dissolved in the water. If studies show that all three phases are bioavailable, then compliance must be based on the total concentration of pesticide in whole water. Similarly, if no data are available regarding bioavailable phases for a given pesticide, then compliance must be based on the total concentration in whole water.
(2) If study results establish that fewer than three phases are bioavailable, then compliance may be based on the concentrations in the bioavailable phases. The most direct way to determine compliance in this case is to measure the concentrations in each phase and determine the total bioavailable concentration. Alternatively, the concentration in the dissolved phase may be estimated from measurement of the total concentration by using the following three-phase equilibrium partitioning model (Chin and Gschwend 1992):

$$C_{\text{dissolved}} = \frac{C_{\text{total}}}{1 + ((K_{\text{OC}} \cdot [\text{SS}])/f_{\text{oc}}) + (K_{\text{DOC}} \cdot [\text{DOC}])} \quad (36)$$

where

$C_{\text{dissolved}}$ = concentration of chemical in dissolved phase (μg/L);
C_{total} = total concentration of chemical in water (μg/L);
K_{OC} = OC–water partition coefficient (L/kg);
[SS] = concentration of suspended solids in water (kg/L);
f_{oc} = fraction of OC in suspended sediment in water;
[DOC] = concentration of DOC in water (kg/L);
K_{DOC} = organic carbon–water partition coefficient (L/kg) for DOC.

The use of this model requires measuring the concentrations of the total pesticide in water, the total suspended solids, and the DOC. Site-specific K_{OC} and K_{DOC} values must also be available.
(3) To estimate the bioavailable concentrations of pesticide without specific knowledge of which phases are bioavailable, passive sampling devices may be used. However, they have a number of technical limitations and will not be useful for determining compliance with acute criteria.

9.5.2 Mixtures

As recommended in TenBrook et al. (2009), only the additive concentration addition model (for pesticides with similar modes of action; Plackett and Hewlett 1952) and the non-additive interaction model (for chemicals that display antagonistic or synergistic interactions; Finney 1942) are included in the UCDM. Two approaches for using the concentration addition model are presented. The non-additive interaction model is presented with the caveat that it can only be applied when a valid coefficient of interaction (K) is available (either a multispecies K value, or individual species K values). Without multispecies K values, this technique should not be used to assess compliance with water quality criteria, but K values for individual species could be used to assess the potential harm from non-additive toxicity on a species by species basis. A final caveat is that application of all of these mixture models requires that each pesticide that is considered in the model has a numeric water quality criterion.

A. Concentration Addition – For Pesticides with Similar Modes of Action

Two equally valid approaches to compliance determination for mixtures of similarly acting pesticides are presented: the toxic unit approach and the RPF approach (as suggested by Felsot 2005). Regulators may choose which to use.

Toxic Unit Approach

According to the toxic unit approach (CVRWQCB 2004), compliance with water quality criteria is determined as follows:

$$\sum_{i=1}^{n} \frac{C_i}{O_i} < 1.0 \tag{37}$$

where

C_i = concentration of toxicant i in water;
O_i = water quality objective/criterion for toxicant i.

As long as the sum is <1.0, the water body is considered to be in compliance with respect to the mixture.

Relative Potency Factor (RPF) Approach

The RPF approach, suggested by Felsot (2005), is analogous to the TEF approach used in assessing toxicity of dioxin and dioxin-like compounds (Van Den Berg et al. 1998). To use this method for a group of similarly acting chemicals, select one chemical (usually the most toxic one) to be the reference chemical. For each chemical in the group, determine an RPF using the following equation:

$$\text{RPF}_i = \frac{\text{Criterion}_{xR}}{\text{Criterion}_{xi}} \tag{38}$$

where

RPF_i = relative potency factor;
Criterion_{xR} = water quality criterion (acute or chronic) of reference chemical (μg/L);
Criterion_{xi} = water quality criterion (acute or chronic) of the ith chemical (μg/L).

Use each RPF value to calculate the toxic equivalents of each component of the mixture with respect to the reference chemical:

$$\text{TE}_i = \text{RPF}_i * C_i \tag{39}$$

where

TE_i = toxic equivalents of ith component of the mixture (μg/L);
RPF_i = relative potency factor of the ith component of the mixture;
C_i = concentration of the ith component of the mixture (μg/L).

Determine compliance with the criterion for the reference chemical using the following equation:

$$\text{TE}_{\text{total}} = C_R + \sum_{n}^{i} \text{TE}_i \tag{40}$$

where

TE_{total} = total toxic equivalents of mixture (μg/L);
C_R = concentration of reference chemical (μg/L).

If $\text{TE}_{\text{total}} \leqq$ the criterion for the reference compound, then the water body is in compliance.

B. Non-additivity: Synergism and Antagonism

If a valid, multispecies interaction coefficient (K; discussed in Section 4.2.2) is available for a known synergist or antagonist over a range of concentrations, then this procedure may be followed to determine compliance of mixtures.

First, determine the adjusted, or effective, concentration of a chemical in the presence of an antagonist or synergist:

$$C_a = C_m(K) \tag{41}$$

where

> C_a = adjusted, or effective, concentration of chemical;
> C_m = concentration measured;
> K = coefficient of interaction, specific to the synergist/antagonist at a particular concentration.

To determine compliance, the adjusted concentration should be compared to the criterion; if the adjusted concentration is higher than the criterion, then the mixture would not be in compliance. Additionally, the adjusted concentration can be used in the additivity models described in Section 9.5.2A. If single-species K values are available over a range of concentrations, this approach may be used to assess the potential for harm, but should not generally be used to determine compliance with criteria. However, if the available single-species K values are for one of the most sensitive species in a data set, then this approach may be used to assess compliance.

For mixtures containing both synergists and antagonists, or multiple synergists/antagonists, Equation (41) can be modified to include multiple K values (LeBlanc, personal communication 2006):

$$C_a = C_m(K_1 K_2 ... K_n) \qquad (42)$$

where

> C_a and C_m are as defined in Equation (41);
> $K_1, K_2, K_n = K$ values for synergist/antagonist 1, 2...n.

This multiple K value approach should not be used to assess compliance, but may be used to assess research needs.

9.5.3 Temperature, pH, and Other Effects

This procedure (taken directly from USEPA 1985, 2003a) can be used for both acute and chronic data. When enough acceptable data (i.e., rated RR by the UCDM) are available to show that toxicity to two or more species (at least one fish and one invertebrate) is similarly related to a water quality characteristic, the relationship should be validated using analysis of covariance (ANCOVA). The ANCOVA may be done with a computer program, or by the manual procedure outlined below in Section 9.5.3A. If two or more factors affect toxicity, multiple regression analysis should be used. If a quantitative relationship is validated with ANCOVA or multiple regression analysis, then the toxicity values obtained in otherwise acceptable studies, conducted under non-standard conditions, may be translated to toxicity values at standard conditions using the procedures described in Section 9.5.3B. If studies are added to the criteria derivation data set using this procedure, then criteria should be recalculated with the additional data. If relationships between toxicity and a water quality characteristic are quantified, then the water quality criteria can be expressed as a function of the given water quality characteristic and used for criteria compliance (Section 9.5.3C).

A. Manual ANCOVA Procedure

If comparable toxicity values from acceptable studies (rated RR) are available at three or more different values of a water quality characteristic for a given species, then perform a least squares regression of the toxicity values on the corresponding values of the water quality characteristic to obtain the slope, and its 95% confidence limits for each species (USEPA 1985, 2003a). These data should be transformed as necessary to optimize model fits.

To decide whether the data for each species are relevant and reasonable, assess the range and number of the tested values of the water quality characteristic and the degree of agreement within and between species. For example, a slope based on six data points might be of limited value if it is based only on data for a very narrow range of values of the water quality characteristic. A slope based on only three data points, however, might be useful if it is consistent with other information and if the three points cover a broad enough range of the water quality characteristic. If useful slopes are not available for at least one fish and one invertebrate, or if the available slopes are statistically dissimilar, or if too few data are available to adequately define the relationship between acute toxicity and the water quality characteristic, then criteria should not be expressed as an equation, and only results of tests conducted under standard conditions should be used for criteria derivation. If a relationship is established, then results of toxicity tests conducted under non-standard conditions can be translated to standard conditions by following the procedures below in Section 9.5.3B. The translated values can then be added to the criteria derivation data set, so that the criteria can be re-calculated with the additional data.

B. Translating Toxicity Values to Different Water Quality Conditions

Once a relationship between toxicity and a water quality characteristic has been validated using ANCOVA or multiple regression analysis, an equation to translate toxicity values at non-standard conditions to toxicity values at standard conditions, or vice versa, can be derived. There are three steps to derive these equations, which are outlined below in this section.

(1) Normalizing toxicity and water quality values

To normalize the toxicity values, calculate the geometric mean of the available acute or chronic toxicity values for each species, then divide each of these toxicity values by the geometric mean for the corresponding species. This normalizes the toxicity values so that the geometric mean of the normalized toxicity values are 1.0 for each species individually and for any combination of species.

Similarly, the values of the water quality characteristic for each species should be normalized individually using this same procedure. Finally, perform a least squares regression of the normalized values of the water quality characteristic on the normalized toxicity values separately for each species. The resulting slopes and 95% confidence limits will be identical to those obtained above with the non-normalized data, but when the data are plotted, the line of best fit for each individual species will go through the point (1,1) in the center of the graph.

(2) Combining species to obtain a pooled slope

Treat all of the normalized data as if they were all for the same species and perform a least squares regression of all normalized acute values on the corresponding normalized values of the water quality characteristic to obtain the pooled acute slope, V, and its 95% confidence limits. The line of best fit for the normalized data set will go through the point (1,1) in the center of the graph.

(3) Deriving the equation to translate toxicity values to different water quality conditions

For each species, calculate the geometric mean of the non-normalized toxicity values (W) and the geometric mean of the values of the non-normalized water quality characteristic (X).

For each species, calculate the mean toxicity value (Y) at a selected value of the water quality characteristic (Z) using the equation:

$$Y = W - V(X - Z) \qquad (43)$$

where

$V =$ pooled slope of the regression curve;
$W =$ geometric mean of toxicity values for a species (at all levels of the water quality characteristic);
$X =$ geometric mean of non-normalized water quality characteristic values for a species;
$Y =$ mean toxicity value for a species at selected water quality characteristic value (Z);
$Z =$ selected value of water quality characteristic.

If data were transformed prior to derivation of regression slopes, then Equation (43) will be

$$\ln Y = \ln W - V(\ln X - \ln Z) \qquad (44)$$

and the toxicity value is calculated as

$$e^Y \qquad (45)$$

Note: alternatively, the toxicity values at Z can be obtained by using Equation (43) or Equations (44) and (45) to adjust each value individually to Z (as opposed to adjusting the geometric mean values), and then calculating the mean of the adjusted values for each species. This alternative procedure permits an examination of the range of the adjusted acute toxicity values for each species.

C. Deriving Criteria at a Selected Level of a Water Quality Characteristic

To determine criteria compliance at a selected level of a given water quality characteristic (Z), all toxicity values in the data set should be translated to Z, using the procedures described in Section 9.5.3B. Next, these translated toxicity values should be used to derive criteria, as described in Sections 9.3 and 9.4, to give new acute and chronic criteria at Z. These translated criteria (A) can then be input to Equations (46) and (47) to determine criteria compliance.

The acute criterion is expressed as

$$\frac{e^{(V[\ln(\text{waterqualitycharacteristic})] + \ln A - V[\ln Z])}}{2} \qquad (46)$$

and the chronic criterion is expressed as

$$e^{(V[\ln(\text{waterqualitycharacteristic})] + \ln A - V[\ln Z])} \qquad (47)$$

where

$V =$ pooled acute slope;
$A =$ acute or chronic criterion at Z derived from SSD, AF, or ACR procedures;
$Z =$ selected value of water quality characteristic.

Because V, A, and Z are known, criteria can be calculated for any selected value of the water quality characteristic.

9.6 Checking Criteria Against Ecotoxicity Data

Criteria derived using the procedures in Sections 9.3 and 9.4 must be evaluated to ensure that they are set at levels that will protect against adverse effects to (1) particularly sensitive species, (2) ecosystems, and (3) TES. If evidence suggests that the 5th percentile will not be protective, criteria may be adjusted downward. The recommended means of making such an adjustment is to use either a lower 95% confidence limit estimate of the 5th percentile (see discussion in Section 3.1.3), or a median or 95% confidence limit estimate of the 1st percentile.

9.6.1 Sensitive Species

Derived criteria should be compared to studies of the most sensitive species to ensure that these species will be protected. If a calculated criterion is higher than the toxicity values reported for a particularly sensitive species, then the criterion may require downward adjustment. This evaluation should be based only on measured toxicity values from acceptable and supplemental studies (i.e., those rated RR, RL, LR, or LL).

9.6.2 Ecosystem and Other Studies

It is necessary to evaluate the criteria against laboratory, field or semi-field data from acceptable multispecies studies (rated R or L) to judge whether they will be protective of ecosystems. To evaluate protection of ecosystems, the derived criteria should be compared to reported ecosystem NOEC values, or to NOEC, EC, IC, or LC values reported for individual species within the system. If toxicity values obtained for appropriate endpoints (i.e., those related to survival, growth, or reproduction) in these studies are lower than the derived criteria, then criteria may be adjusted downward. Adjustment of criteria upward is not recommended because single-species data have indicated this concentration to be protective and increasing the criteria may result in toxicity to sensitive species.

9.6.3 Threatened and Endangered Species

Criteria derived to protect the most sensitive species in ecosystems should also be protective of TES. However, a few tools are available to investigate this more rigorously. The guidance presented here may be used to assess whether criteria derived by the UCDM will be protective of TES.

First, obtain the latest list of the appropriate federal or state TES. The California TES list is available from the CDFG web site (www.dfg.ca.gov/hcpb/species/t_e_spp/tespp.shtml; CDFG 2006a, b).

Then, for comparison to acute criteria:

(1) Compare criteria to toxicity values from acceptable studies of effects on TES.
(2) If no toxicity values are available for a TES, but an acceptable acute toxicity value is available for a surrogate species in the same family or genus as the TES, then use the ICE program (v. 1.0; available at http://www.epa.gov/ceampubl/fchain/index.htm) to estimate a toxicity value for the TES (Asfaw et al. 2003). Compare this estimated value to the acute criterion.
(3) If no values for surrogate species are available, and if the chemical of interest has a narcotic mode of action, select a QSAR (e.g., from OECD 1995a; RIVM 2001) that can be used to estimate toxicity to the TES, or to a surrogate species, based on a log K_{OW} value. Note that while many industrial chemicals have a narcotic mode of action, very few pesticides fall into this category. Fumigants (e.g., methyl bromide, naphthalene, chloropicrin, and others) are a class of pesticides with a narcotic mode of action (USEPA 2006).

Table 25 QSARs for estimating toxicity from K_{OW} for chemicals acting by narcosis; from OECD (1995a) and RIVM (2001)

Species	Equation
Acute Toxicity	Summarized in OECD 1995a
Pimephales promelas	log LC$_{50}$ (mM) = $-$ 0.94 log K_{OW} + 0.94 log (0.00068 K_{OW} + 1) + 1.75 (Veith et al. 1983)
Poecilia reticulata	log LC$_{50}$ (mM) = $-$0.87 log K_{OW} + 1.87 (Konemann 1981)
Daphnia magna	log EC$_{50}$ (mM) = $-$ 0.91 log K_{OW} + 1.72 (Hermens et al. 1984)
Chronic Toxicity	Summarized in OECD (1995a)
Brachydanio rerio/Pimephales promelas	log NOEC (mM) = $-$ 0.90 log K_{OW} + 0.8 (Call et al. 1985; Van Leeuwen et al. 1990)
Daphnia magna	log NOEC (mM) = $-$ 1.04 log K_{OW} + 1.30 (Dewolf et al. 1988; Kuhn et al. 1989)
Daphnia magna	log NOEC (mM) = $-$ 1.07 log K_{OW} + 1.25 (Dewolf et al. 1988)
Selenastrum capricornutum	log NOEC (mM) = $-$ 1.00 log K_{OW} + 1.77 (Calamari et al. 1983; Galassi et al. 1988)
Chronic Toxicity	Summarized in RIVM (2001) from Van Leeuwen et al. (1992), Verhaar et al. (1994)
Skeletonema costacum	log NOEC (M) = $-$ 0.72 log K_{OW} $-$ 1.42
Scenedesmus subspicatus	log NOEC (M) = $-$ 0.86 log K_{OW} $-$ 1.41
Selenastrum capricornutum	log NOEC (M) = $-$ 1.00 log K_{OW} $-$1.71
Tetrahymena pyriformis	log NOEC (M) = $-$ 0.80 log K_{OW} $-$ .128
Lymnaea stagnalis	log NOEC (M) = $-$ 0.86 log K_{OW} $-$ 2.08
Nitocra spinipes	log NOEC (M) = $-$ 0.78 log K_{OW} $-$ 2.14
Daphnia magna	log NOEC (M) = $-$ 1.04 log K_{OW} $-$ 1.70
Aedes aegypti	log NOEC (M) = $-$ 1.09 log K_{OW} $-$ 1.36
Culex pipiens	log NOEC (M) = $-$ 0.86 log K_{OW} $-$ 1.98
Brachydanio rerio/Pimephales promelas	log NOEC (M) = $-$ 0.87 log K_{OW} $-$ 2.35
Ambystoma mexicanum	log NOEC (M) = $-$ 0.88 log K_{OW} $-$ 1.89
Rana temporaria	log NOEC (M) = $-$ 1.09 log K_{OW} $-$ 1.47
Xenopus laevis	log NOEC (M) = $-$ 0.90 log K_{OW} $-$ 1.79

For comparison to chronic criteria:

(1) Compare criteria to toxicity values from acceptable studies of the effects on TES.
(2) If no values for surrogate species are available, and if the chemical of interest has a narcotic mode of action, select a QSAR (e.g., from OECD 1995a; RIVM 2001) that can be used to estimate toxicity to the TES, or to a surrogate species, based on a log K_{OW} value.

The QSARs from RIVM (2001) and OECD (1995a) are given in Table 25. These are presented as examples and do not preclude the use of other QSARs that may be established in future published studies.

If no data for TES or acceptable surrogates are available, and if no applicable QSARs are available, then no special methods are available to assess whether the criteria will be protective of these species; however, protection of TES is expected since criteria are derived to protect all species. If any of the above comparisons reveal that a criterion is higher than any of the TES toxicity values (or estimated toxicity values), then the criterion may need downward adjustment.

9.7 Partitioning to Other Environmental Compartments

The criteria should also be checked to determine if they are in conflict with any existing guidelines for (1) wildlife and human health due to bioaccumulation and/or (2) other environmental compartments due to partitioning of chemicals from the water compartment. Information that indicates the criteria may be in conflict with other protection goals should be flagged for further review by environmental managers. Results of these sections should not be used to alter final criteria.

9.7.1 Bioaccumulation/Secondary Poisoning

For bioaccumulative chemicals it is important to be sure that water quality criteria are set at levels that do not lead to unacceptable levels of chemicals in food items. This section presents a procedure for evaluating the chronic criteria based on the possibility of secondary poisoning of wildlife, or possible human health effects, which may result from consumption of fish or other food items that may have bioaccumulated the pesticide of concern. Acute criteria do not require this evaluation because they are intended to protect against short periods of elevated pesticide concentrations, making the equilibrium model inappropriate. To perform this evaluation for wildlife, studies that demonstrate adverse effects from dietary intake of toxicants are used; for human health, USFDA action limits for the chemical of concern are necessary.

The first step is to determine if the chemical of interest is known to bioaccumulate, or has the potential to bioaccumulate. Candidates are chemicals that have been shown to bioaccumulate in well-conducted studies (i.e., consistent with standard methods), or have one or more of the following characteristics: log K_{OW} >3, (ECB 2003; OECD 1995a), molecular weight <1,000 (OECD 1995a), molecular diameter <5.5 Å (OECD 1995a), molecular length <5.5 nm (OECD 1995a), solid–water partition coefficient (log K_d) >3; highly adsorbent (ECB 2003), or belong to a class of chemicals that are known to be bioaccumulative (ECB 2003). Chemicals are not expected to bioaccumulate if they are reactive and/or readily metabolized.

The next steps only apply if a chemical is bioaccumulative, or has the potential to bioaccumulate, and if wildlife dietary toxicity data or USFDA action levels are available. To evaluate effects on humans, the USFDA action levels for fish tissues

Table 26 Chemical names of pesticides and other chemicals mentioned in text

Common name	CAS number	Chemical name
2,4,5-trichlorophenol	95-95-4	2,4,5-trichlorophenol
3-trifluoromethyl-4-nitrophenol (TFM)	88-30-2	α,α,α-tirfluoro-4-nitro-m-cresol, sodium salt
3,4-dichlorophenol	95-77-2	3,4-dichlorophenol
4-chloroaniline	106-47-8	p-chloroaniline
4-nitrophenol	100-02-7	4-nitrophenol
4-nonylphenol	104-40-5	4-nonylphenol
17α-ethinylestradiol	57-63-6	17α-ethinylestradiol
Aldrin	309-00-2	(1R,4S,4aS,5S,8R,8aR)-rel-1,2,3,4,10,10-hexachloro-1,4,4a,5,8,8a-hexahydro-1,4:5,8-dimethanonaphthalene
Antimycin	642-15-9	(2{R},3{S},6{S},7{R},8{R})-3-[(3-formamido-2-hydroxybenzoyl)amino]-8-hexyl-2,6-dimethyl-4,9-dioxo-1,5-dioxonan-7-yl 3-methylbutanoate
Atrazine	1912-24-9	6-chloro-N-ethyl-N'-(1-methylethyl)-1,3,5-triazine-2,4-diamine
Azinphos-methyl	86-50-0	O,O-dimethyl S-[(4-oxo-1,2,3-benzotriazin-3(4H)-yl)methyl] phosphorodithioate
Bayluscid	50-65-7	5-chloro-N-(2-chloro-4-nitrophenyl)-2-hydroxybenzamide
Bifenthrin	82657-04-3	(2-methyl[1,1'-biphenyl]-3-yl)methyl (1R,3R)-rel-3-[(1Z)-2-chloro-3,3,3-trifluoro-1-propenyl]-2,2-dimethylcyclopropanecarboxylate
Carbaryl	63-25-2	1-naphthalenyl methylcarbamate
Carbofuran	1563-66-2	2,3-dihydro-2,2-dimethyl-7-benzofuranyl methylcarbamate
Chlordane	57-74-9	1,2,4,5,6,7,8,8-octachloro-2,3,3a,4,7,7a-hexahydro-4,7-methano-1H-indene
Chlorobenzuron	57160-47-1	2-chloro-N-[(4-chlorophenyl)carbamoyl]benzamide
Chloropicrin	76-06-2	trichloronitromethane
Chlorpyrifos	2921-88-2	O,O-diethyl O-(3,5,6-trichloro-2-pyridinyl) phosphorothioate
DDE	72-55-9	1,1-bis-(4-chlorophenyl)-2,2-dichloroethene
DDT	50-29-3	1,1'-(2,2,2-trichloroethylidene)bis[4-chlorobenzene]
Diazinon	333-41-5	O,O-diethyl O-[6-methyl-2-(1-methylethyl)-4-pyrimidinyl] phosphorothioate
Dieldrin	60-57-1	(1aR,2R,2aS,3S,6R,6aR,7S,7aS)-rel-3,4,5,6,9,9-hexachloro-1a,2,2a,3,6,6a,7,7a-octahydro-2,7:3,6-dimethanonaphth[2,3-b]oxirene

Table 26 (continued)

Common name	CAS number	Chemical name
Dioxin	290-67-5	1,4-dioxin
Endosulfan	115-29-7	6,7,8,9,10,10-hexachloro-1,5,5a,6,9,9a-hexahydro-6,9-methano-2,4,3-benzodioxathiepin 3-oxide
Endrin	72-20-8	(1aR,2R,2aR,3R,6S,6aS,7S,7aS)-rel-3,4,5,6,9,9-hexachloro-1a,2,2a,3,6,6a,7,7a-octahydro-2,7:3,6-dimethanonaphth[2,3-b]oxirene
Esfenvalerate	51630-58-1	(S)-cyano(3-phenoxyphenyl)methyl (αS)-4-chloro-α-(1-methylethyl)benzeneacetate
Fenitrothion	122-14-5	O,O-dimethyl O-(3-methyl-4-nitrophenyl) phosphorothioate
Fenvalerate	51630-58-1	cyano(3-phenoxyphenyl)methyl 4-chloro-α-(1-methylethyl)benzeneacetate
Fluoranthene	206-44-0	Fluoranthene
Glyphosate	1071-83-6	N-(phosphonomethyl)glycine
Heptachlor	76-44-8	1,4,5,6,7,8,8-heptachloro-3a,4,7,7a-tetrahydro-4,7-methano-1H-indene
Lambda-cyhalothrin	91465-08-6	(R)-cyano(3-phenoxyphenyl)methyl (1S,3S)-rel-3-[(1Z)-2-chloro-3,3,3-trifluoro-1-propenyl]-2,2-dimethylcyclopropanecarboxylate
Lindane	58-89-9	(1α,2α,3β,4α,5α,6β)-1,2,3,4,5,6-hexachlorocyclohexane
Malathion	121-75-5	diethyl [(dimethoxyphosphinothioyl)thio]butanedioate
Methoxychlor	72-43-5	1,1'-(2,2,2-trichloroethylidene)bis[4-methoxybenzene]
Methyl bromide	74-83-9	Bromomethane
Naphthalene	91-20-3	Naphthalene
o-toluidine	95-53-4	2-amino-1-methylbenzene
Parathion	56-38-2	O,O-diethyl O-(4-nitrophenyl) phosphorothioate
Pentachlorophenol	87-86-5	Pentachlorophenol
Permethrin	52645-53-1	(3-phenoxyphenyl)methyl 3-(2,2-dichloroethenyl)-2,2-dimethylcyclopropanecarboxylate
Phenanthrene	85-01-8	Phenanthrene
Piperonyl butoxide (PBO)	51-03-6	5-[[2-(2-butoxyethoxy)ethoxy]methyl]-6-propyl-1,3-benzodioxole
Pirimiphos-methyl	29232-93-7	O-[2-(diethylamino)-6-methyl-4-pyrimidinyl] O,O-dimethyl phosphorothioate
Rotenone	83-79-4	(2R,6aS,12aS)-1,2,12,12a-tetrahydro-8,9-dimethoxy-2-(1-methylethenyl)[1]benzopyrano[3,4-b]furo[2,3-h][1]benzopyran-6(6aH)-one

Table 26 (continued)

Common name	CAS number	Chemical name
Temephos	3383-96-8	O,O'-(thiodi-4,1-phenylene) bis(O,O-dimethyl phosphorothioate)
Tetrabromobisphenol-A	79-94-7	2,6-dibromo-4-[1-(3,5-dibromo-4-hydroxy-phenyl)-1-methyl-ethyl]phenol
Toxaphene	8001-35-2	1,4,5,6,7,7-hexachloro-3-(dichloromethyl)-2,2-dimethyl-norbornane
Tributyltin	688-73-3	tributylstannane

should be obtained. To evaluate effects on wildlife, one must obtain and rate toxicity studies from wildlife that have significant food sources in water. Often, Mallard duck toxicity values are generated for pesticide registration and are available from the USEPA (Table 3). Toxicity values from wildlife studies that are rated R or L (Table 12) may be used for this evaluation. A chronic NOEC is the preferred wildlife toxicity value to use in this section, but sub-acute toxicity values may be used if a NOEC is not available. Three common oral wildlife toxicity values that may be useful are described below:

(1) Acute (LC_{50}): one time dose, usually force fed (oral gavage/intubation), and the toxicity value is reported as mg/kg body wt. Since this value is expressed per body weight, rather than as a feed concentration, it is not recommended for use in this section.
(2) Sub-acute (LC_{50}): in which the compound is administered in the feed to the animals for 2 weeks to several months, and the toxicity value is usually reported as ppm concentration in feed.
(3) Chronic (NOEC and LOEC): similar to the exposure conditions in a sub-acute study, but effects on reproduction parameters are monitored.

Measured (preferred) or estimated BCF, BMF, and/or BAF values for food items are required for the calculation. The following equations can be used to translate dietary NOEC or LC_{50} values, or USFDA action levels, into water NOEC values (adapted from ECB 2003):

$$\text{NOEC}_{\text{water}} = \frac{\text{NOEC}_{\text{oral\_predator}}}{\text{BCF}_{\text{food\_item}} \times \text{BMF}_{\text{food\_item}}} \quad (48)$$

or

$$\text{NOEC}_{\text{water}} = \frac{\text{LC}_{50,\text{oral-predator}}}{\text{BCF}_{\text{food\_item}} \times \text{BMF}_{\text{food\_item}}} \quad (49)$$

where

>NOEC$_{water}$ = NOEC in water; concentration in water below this level is not expected to lead to bioaccumulation to harmful levels in food items;
>NOEC$_{oral\_predator}$ = dietary NOEC value for wildlife or USFDA action level (mg pesticide/kg food);
>LC$_{50,oral\_predator}$ = dietary LC$_{50}$ value for wildlife (mg pesticide/kg food);
>BCF$_{food\_item}$ = bioconcentration factor; ratio of concentration of chemical in tissue of food item due to water-only exposure to concentration in water; whole-body, wet-wt value (ECB 2003; OECD 1995a; USEPA 1985, 2003a),
>BMF$_{food\_item}$ = biomagnification factor in food item; ratio of concentration of chemical in predator to concentration in prey items; lipid-normalized, if possible (ECB 2003).

If no measured BCF is available, a value can be estimated using the log K_{OW} from the following linear free energy relationship (Mackay 1982), which was derived for chemicals with log K_{OW} values ranging from ~2 to ~7:

$$\log \text{BCF} = \log K_{OW} - 1.32 \quad (50)$$

Crosby (1998) cautions that predictions that rely on this equation are less accurate for compounds with log BCF values above 5 or below 2. If Equation (50) gives a result outside this range, then a more appropriate linear free energy relationship should be sought in the literature.

If no measured BMF is available, use an appropriate default value from Table 22 (based on log K_{OW} or BCF; ECB 2003). Note that the default BMF values that are based on log K_{OW} in Table 22 represent high estimates from studies showing no biomagnification of compounds with log K_{OW} values <6 (Berglund et al. 2000; Varó et al. 2002). In the case of chlorpyrifos (log K_{OW} = 4.96), Varó et al. (2002) attribute the lack of biomagnification, in part, to the biotransformation and depuration ability of organisms at higher trophic levels. For compounds that are readily biotransformed, the default values that are based on BCF should be used in favor of those based on log K_{OW}.

Alternatively, if a BAF is available for fish, then Equation (48) is modified to

$$\text{NOEC}_{water} = \frac{\text{NOEC}_{oral\_predator}}{\text{BAF}_{fish}} \quad (51)$$

where

>NOEC$_{water}$ = NOEC in water;
>NOEC$_{oral\_predator}$ = dietary NOEC for wildlife or USFDA action level (mg pesticide/kg food);

BAF_{fish} = bioaccumulation factor in fish; ratio of concentration of chemical in tissue due to water plus dietary exposure to concentrations in water; lipid-normalized for chemicals with log K_{OW} >3.

Equation (49) can be modified in the same way, substituting BAF for (BCF*BMF).

If no BAF value is available, then Equation (48) or Equation (49) must be used, and if no measured BMF value is available, then the appropriate default value should be used (Table 22). If multiple BCF, BAF, or BMF values are available for a chemical, the geometric mean of all acceptable values should be used.

To determine compliance, compare the $NOEC_{water}$ derived from one of the equations in this section to the chronic water quality criterion. If it is above the criterion, then no adjustment of the criterion is necessary. If the $NOEC_{water}$ is below the criterion, then indicate in the final criteria statement that these criteria may not be protective of all beneficial uses based on the bioaccumulation/secondary poisoning section and that additional review is needed. Discussion of such additional review is beyond the scope of the UCDM.

9.7.2 Harmonization with Air or Sediment Criteria

Pesticides in the water may sorb to sediment or volatilize into the air and cause toxicity to organisms in those compartments. Steady-state environmental models may be used to assess harmony, or coherence, of chronic criteria across all environmental media. It is only necessary to consider chronic criteria in this analysis because it is based on equilibrium partitioning, which is not appropriate to apply to acute criteria. If there are no levels of concern established for sediment, air, or biota compartments, then there is no need to use this procedure. Concern for bioaccumulation/secondary poisoning that may affect wildlife or human health is addressed by the procedure outlined in Section 9.7.1.

Several acceptable, freely available models are available to assess coherence of chronic criteria across all environmental media, including the following:

(1) Exposure Analysis Modeling System (EXAMS; Burns 2004) available from the USEPA Center for Exposure Assessment Modeling (CEAM; http://www.epa.gov/ceampubl/swater/index.htm). The software can be downloaded directly from the USEPA website.
(2) MacKay's Fugacity-Based Environmental Equilibrium Partitioning Models (Mackay 2001), from the Canadian Environmental Monitoring Center (CEMC; http://www.trentu.ca/cemc/). The software can be downloaded directly from the CEMC website.

The different fate models vary in complexity, and require the use of default environmental parameter values when measured values are not available, but they can provide rough estimates of equilibrium concentrations of chemicals in all environmental compartments based on a given concentration in water (i.e., the chronic criterion concentration) and a few physical–chemical parameters for the chemical.

Because of its relative ease of use, the Level I fugacity model is recommended as a rough first-pass evaluation of equilibrium concentrations. In using this model, the total mass of chemical in the system is adjusted until the equilibrium concentration in water is at the chronic criterion level. The model should be run over a range of values for parameters that may affect equilibria (e.g., OC levels or fish lipid levels). If no harmonization problems are apparent from a series of Level I analyses (i.e., steady-state concentrations in all compartments are below their respective levels of concern), then no further analysis is necessary. However, if any problems are identified, then site-specific data should be obtained to allow more refined modeling.

For all models used in this analysis, it is important to state all input parameters, conditions, and assumptions. The model outputs can then be compared to appropriate levels of concern established for the non-water compartments (e.g., sediment or air quality criteria or USFDA action levels). If the steady-state concentrations in all compartments are acceptable, then the water quality criterion is acceptable. If the concentration in a non-water compartment is projected to exceed a concentration of concern, then indicate, in the final criteria statement, that these criteria may not be protective of all beneficial uses (based on the harmonization with air or sediment criteria section), and that additional review is needed.

9.8 Reviewing Assumptions and Limitations to Derived Criteria

The assumptions, limitations, and uncertainties involved in criteria generation should be available to inform environmental managers of the accuracy and confidence in criteria. In Sections 1, 2, 3, 4, 5, 6, 7, and 8 these points are individually discussed as each procedure is presented. It is recommended that researchers should summarize any data limitations that may affect the procedure used in determining the final criteria in a separate section of the final criteria report. The final criteria statement (Section 9.9) should also briefly review these points to make it obvious how the final criterion was derived. An example of an important limitation that would affect the derivation process would be missing taxa in the data set that required use of an AF instead of an SSD. The calculations of different distributional estimates (median and lower 95% confidence limits of the 5th and 1st percentiles) included in Section 9.3.2A may be used to evaluate possible uncertainty in the resulting criteria. If ecotoxicity data from sensitive species, TES, or multispecies studies (Section 9.6) indicate that the median 5th percentile estimate is likely to be underprotective, a lower distributional estimate (1st percentile or lower 95% confidence limit) may be used in the criteria calculations. The consideration of water quality effects and sensitive ecotoxicity data (Sections 9.5 and 9.6) may indicate data limitations that should be discussed in this section of the criteria report. Moreover, criteria reports should be periodically reviewed so that they can reflect the most recent published literature.

9.9 Final Criteria Statement

The procedures used to calculate criteria should be briefly summarized, with any caveats to the criteria highlighted. Final criteria statements should briefly review any other considerations (from Section 9.6) that may be important for policy makers to consider.

Criteria are properly stated as follows (USEPA 1985, 2003a):

Aquatic life should not be affected unacceptably if the 4-day average concentration of (1) does not exceed (2) μg/L more than once every 3 years on the average and if the 1-h average concentration does not exceed (3) μg/L more than once every 3 years on average.

where

(1) = insert name of chemical;
(2) = insert the chronic criterion;
(3) = insert the acute criterion.

These averaging periods and the frequency of exceedance may be modified if data and/or models become available that can scientifically defend altering them.

10 Summary

Water quality criteria are used by environmental regulators to determine safe levels of contaminants for the protection of aquatic life. The USEPA established a methodology for deriving water quality criteria in 1985 (USEPA 1985), but this methodology no longer reflects the most recent research results in aquatic ecotoxicology and environmental risk assessment. The recent review by TenBrook et al. (2009) identified relevant aspects of criteria derivation methodologies, and compared methodologies from several countries. In this review we elaborate on the work of TenBrook et al. (2009) by selecting procedures from those methods they identified to form a new criteria derivation methodology. This new methodology, the UCDM, is a compilation of procedures for each step of the criteria derivation process, which are designed to be flexible and suitable for use with a wide variety of data sets. In this review we address each aspect of and the rationale for selecting each component of the UCDM. We include a section written in the form of a standard operating procedure for streamlined use of the UCDM for deriving pesticide criteria.

The UCDM, and all other existing methodologies examined by TenBrook et al. (2009), use effects data to derive aquatic life criteria; thus, all methods are limited by availability of relevant toxicity data. One of the major limitations of the USEPA (1985) methodology is that criteria cannot be calculated when there are less than eight toxicity values, which represent eight different taxonomic categories, for a given pesticide. The pesticide registration process requires the generation of some aquatic life toxicity data, but registrants are rarely, if ever, required to provide toxicity data that would satisfy all eight of the data requirements for criteria derivation by the USEPA (1985) methodology. The USEPA (1985) criteria derivation methodology does not provide guidance for data sets that do not meet the data requirements; therefore, one goal of the UCDM was to create a methodology that provided criteria derivation guidance for limited data sets. Water quality criteria can be generated using the UCDM for pesticides with data sets as small as one datum, although more data is preferred and will result in more robust criteria.

Because toxicity data constitute the core of any criteria derivation methodology, the UCDM emphasizes data quality, as well as data quantity. A detailed numeric rating scheme was developed to make data evaluation a more transparent and objective process. However, this numeric system does not negate the need for familiarity with standard test methods, the principles of toxicology, and accepted laboratory practices to evaluate available studies. Data collection, evaluation, and reduction are the first steps of criteria derivation, according to the UCDM, and specific procedures are given for each step, with the goal of assembling a large, diverse, and high-quality data set for a given pesticide.

Once a data set is assembled, the UCDM includes several options for criteria calculation, which depend on the size and diversity of the data set. The SSD approach utilized in the Australia/New Zealand criteria derivation guidelines (ANZECC and ARMCANZ 2000) was selected for use in the UCDM, but with the modification that it may be used to derive acute or chronic criteria, when at least five toxicity data values, representing five different families, are available. To arrive at the acute criterion (using the SSD approach), the 5th percentile values that are derived from acute SSDs are divided by two, whereas 5th percentile values derived from chronic data becomes the chronic criterion. For data sets consisting of more than eight toxicity values, the Burr Type III SSD is recommended, whereas the log-logistic SSD is recommended for use with data sets consisting of 5–8 toxicity values, to avoid overfitting limited data sets. For acute data sets that do not meet the five taxonomic requirements of the SSD approach, an AF procedure similar to that used in the Great Lakes methodology (USEPA 2003a) is included, but new factors, specific to pesticides, have been developed. When fewer than five chronic data values are available, chronic criteria are to be derived from acute criteria that are divided by an ACR. When data are available, it is recommended that ACRs are calculated with effects data. A default ACR, which was calculated from existing high-quality pesticide data sets, is provided for use with limited chronic data sets.

Criteria are stated in terms of magnitude, duration, and frequency. For acute criteria a 1-h averaging period was established, while for chronic criteria a 4-day averaging period was established. For both acute and chronic criteria, an allowable frequency of criteria exceedance of once-in-three-years was established. The duration and frequency components of the UCDM criteria are equivalent to those used by the USEPA (1985) methodology.

According to the UCDM, criteria are derived using data solely from single-species toxicity tests with aqueous exposures. However, to ensure that criteria will be protective when applied to more complex environmental conditions, guidance is given in the UCDM for how to evaluate the criteria against several other types of data. If the effects of bioavailability, pesticide mixtures, and/or water quality (e.g., pH, temperature) are quantifiable, those effects are incorporated into criteria compliance. Procedures are also presented for evaluating if criteria are likely to be protective based on study results of particularly sensitive species, threatened and endangered species, and multispecies ecosystems. Although it is generally outside of the realm of protection of aquatic life, the potential for bioaccumulation in wildlife and humans, and partitioning into other environmental media are addressed in the UCDM. These aspects of criteria evaluation give a more complete picture of the complex ecosystems that environmental regulators are commissioned to protect and manage.

The UCDM is a water quality criteria derivation methodology, specifically designed for pesticides, that incorporates the most recent research in aquatic toxicology and environmental risk assessment. This water quality criteria derivation methodology can be used by environmental managers to calculate appropriate levels of pesticides in water bodies to ensure protection of aquatic life.

Acknowledgments We thank the following reviewers: Daniel McClure (CVRWQCB), Joshua Grover (CVRWQCB), Zhimin Lu (CVRWQCB), Lawrence R. Curtis (Oregon State University), Brian Finlayson (CDFG), Evan P. Gallagher (University of Washington), John P. Knezovich (Lawrence Livermore National Laboratory), and Marshall Lee (CDPR). This project was funded through a contract with the CVRWQCB. Mention of specific products, policies, or procedures does not represent endorsement by the CVRWQCB. The contents also do not necessarily reflect the views or policies of the USEPA nor does mention of trade names or commercial products constitute endorsement or recommendation for use.

Appendix: Acute Chlorpyrifos Data Collected for Criteria Derivation Using the UCDM Derivation

Final acute toxicity data set for chlorpyrifos. All studies were rated relevant and reliable (RR) and were conducted at standard temperature[a]

Species	Common identifier	Family	Test type	Meas/Nom	Chemical grade (%)	Duration (h)	Temp (°C)	Endpoint	Age/size	LC/EC_{50} (μg/L)	References
Ceriodaphnia dubia	Cladoceran	Daphniidae	S	Meas	99.0	96	25	Mortality	<24 h	0.053	Bailey et al. (1997)
Ceriodaphnia dubia	Cladoceran	Daphniidae	S	Meas	99.0	96	25	Mortality	<24 h	0.055	Bailey et al. (1997)
Ceriodaphnia dubia	Cladoceran	Daphniidae	SR	Meas	99.0	96	24.6	Mortality	<24 h	0.13	CDFG (1992 g)
Ceriodaphnia dubia	Cladoceran	Daphniidae	SR	Meas	99.0	96	24.3	Mortality	<24 h	0.08	CDFG (1992d)
Ceriodaphnia dubia	Cladoceran	Daphniidae	SR	Meas	99.8	96	24.6	Survival	<24 h	0.0396	CDFG (1999)
Ceriodaphnia dubia	Cladoceran	Daphniidae						Geometric mean		**0.0654**	
Chironomus tentans	Insect	Chironomidae	S	Meas	98.0	96	21	Immobility	3–4th instar	0.16	Belden and Lydy (2006)
Chironomus tentans	Insect	Chironomidae	S	Meas	90.0	96	21	Immobility	4th instar	0.17	Lydy and Austin (2005)
Chironomus tentans	Insect	Chironomidae	S	Meas	98.0	96	20	Immobility + mortality	4th instar	0.39	Belden and Lydy (2000)

Appendix (continued)

Species	Common identifier	Family	Test type	Meas/Nom	Chemical grade (%)	Duration (h)	Temp (°C)	Endpoint	Age/size	LC/EC$_{50}$ (μg/L)	References
Chironomus tentans	Insect	Chironomidae						Geometric mean		**0.220**	Harmon et al. (2003)
Daphnia ambigua	Cladoceran	Daphniidae	S	Meas	99.0	48	21	Immobility	Neonates	**0.035**	Kersting and Van Wijngaarden (1992)
Daphnia magna	Cladoceran	Daphniidae	S	Meas	99.0	48	19.5	Mortality	<24 h	1.0	Burgess (1988)
Daphnia magna	Cladoceran	Daphniidae	FT	Nom (most)	95.5	48	18–21	Mortality	<24 h	0.10	
Daphnia magna								Geometric mean		**0.32**	
Daphnia pulex	Cladoceran	Daphniidae	S	Meas	Technical	48	20	Immobility	<24 h	**0.25**	Van Der Hoeven and Gerritsen (1997)
Hyalella azteca	Amphipod	Hyalellidae	S	Meas	90.0	96	20	Mortality	14–21 days	0.0427	Anderson and Lydy (2002)
Hyalella azteca	Amphipod	Hyalellidae	SR	Meas	98.1	96	19	Mortality	14–21 days	0.138	Brown et al. (1997)

Appendix (continued)

Species	Common identifier	Family	Test type	Meas/Nom	Chemical grade (%)	Duration (h)	Temp (°C)	Endpoint	Age/size	LC/EC$_{50}$ (μg/L)	References
Hyalella azteca								Geometric mean		**0.077**	
Ictalurus punctatus	Channel catfish	Ictaluridae	FT	Meas	99.9	96	17.3	Mortality	7.9 g	**806**	Phipps and Holcombe (1985)
Lepomis macrochirus	Bluegill	Centrarchidae	FT	Meas	99.9	96	17.3	Mortality	0.8 g	10	Phipps and Holcombe (1985)
Lepomis macrochirus	Bluegill	Centrarchidae	FT	Meas	99.9	96	22	Mortality	2.1 g	5.8	Bowman (1988)
Lepomis macrochirus								Geometric mean		**7.6**	
Neomysis mercedis	Opossum shrimp	Mysidae	SR	Meas	99.0	96	17.4	Mortality	<5 days	0.15	CDFG (1992f)
Neomysis mercedis	Opossum shrimp	Mysidae	SR	Meas	99.0	96	17.2	Mortality	<5 days	0.16	CDFG (1992b)
Neomysis mercedis	Opossum shrimp	Mysidae	SR	Meas	99.0	96	17.1	Mortality	<5 days	0.14	CDFG (1992e)
Neomysis mercedis	Opossum shrimp	Mysidae						Geometric mean		**0.150**	
Oncorhynchus mykiss	Rainbow trout	Salmonidae	FT	Meas	99.9	96	12	Mortality	Juvenile	8.0	Holcombe et al. (1982)
Oncorhynchus mykiss	Rainbow trout	Salmonidae	FT	Meas	95.9	96	12	Mortality	0.25 g	25.0	Bowman (1988)

Appendix (continued)

Species	Common identifier	Family	Test type	Meas/Nom	Chemical grade (%)	Duration (h)	Temp (°C)	Endpoint	Age/size	LC/EC$_{50}$ (µg/L)	References
Oncorhynchus mykiss								Geometric mean		**14**	
Oncorhynchus tshawytscha	Chinook salmon	Salmonidae	SR	Meas	99.5	96	14.8	Mortality	Juvenile	**15.96**	Wheelock et al. (2005)
Orconectes immunis	Crayfish	Cambaridae	FT	Meas	99.9	96	17.3	Mortality	1.8 g	**6**	Phipps and Holcombe (1985)
Pimephales promelas	Fathead minnow	Cyprinidae	FT	Meas	99.9	96	25	Mortality	32 days	200	Geiger et al. (1988)
Pimephales promelas	Fathead minnow	Cyprinidae	FT	Meas	99.9	96	25	Mortality	31–32 days	203	Holcombe et al. (1982)
Pimephales promelas	Fathead minnow	Cyprinidae	FT	Meas	98.7	96	25	Mortality	Newly hatched	140	Jarvinen and Tanner (1982)
Pimephales promelas	Fathead minnow	Cyprinidae						Geometric mean		**178**	

Appendix (continued)

Species	Common identifier	Family	Test type	Meas/Nom	Chemical grade (%)	Duration (h)	Temp (°C)	Endpoint	Age/size	LC/EC$_{50}$ (μg/L)	References
Procloeon sp.	Insect	Baetidae	SR	Meas	99	48	21.3°C	Mortality	0.5–1.0 cm	0.1791	Anderson et al. (2006)
Procloeon sp.	Insect	Baetidae	SR	Meas	99	48	21.3°C	Mortality	0.5–1.0 cm	0.0704	Anderson et al. (2006)
Procloeon sp.	Insect	Baetidae	SR	Meas	99	48	21.3°C	Mortality	0.5–1.0 cm	0.0798	Anderson et al. (2006)
								Geometric mean		**0.100**	
Pungitius pungitius	Stickleback	Gasterosteidae	FT	Meas	99.8	96	19	Mortality	Adult	**4.7**	Van Wijngaarden et al. (1993)
Simulium vittatum IS-7	Insect	Simuliidae	S	Meas	98.0	24	19	Mortality	2nd and 3rd instar	**0.06**	Hyder et al. (2004)
Xenopus laevis	African clawed frog	Pipidae	SR	Nom	99.80	96	24.7	Mortality	<24 h	**2,410**	El-Merhibi et al. (2004)

Values in bold are species mean acute values. S: static; SR: static renewal; FT: flow-through.
[a]Standard temperatures are particular for each species. See standard methods referenced in Tables 9 and 10.

References

Aldenberg T (1993) ETX 1.3a. A program to calculate confidence limits for hazardous concentrations based on small samples of toxicity data. National Institute of Public Health and the Environment (RIVM), Bilthoven, The Netherlands

Aldenberg T, Slob W (1993) Confidence limits for hazardous concentrations based on logistically distributed NOEC toxicity data. Ecotoxicol Environ Saf 25:48–63

Aldenberg T, Jaworska JS (2000) Uncertainty of the hazardous concentration and fraction affected for normal species sensitivity distributions. Ecotoxicol Environ Saf 46:1–18

Aldenberg T, Luttik R (2002) Extrapolation factors for tiny toxicity data sets from species sensitivity distributions with known standard deviation. In: Posthuma L, Suter IIGW, Traas TP (eds) Species sensitivity distributions in ecotoxicology. Lewis Publishers, New York, NY

Alvarez DA, Petty JD, Huckins JN, Jones-Lepp TL, Getting DT, Goddard JP, Manahan SE (2004) Development of a passive, in situ, integrative sampler for hydrophilic organic contaminants in aquatic environments. Environ Toxicol Chem 23:1640–1648

Anderson JW, Moore LJ, Blaylock JW, Woodruff DL, Keissa SL (1977) Bioavailability of sediment-sorbed naphthalenes to the sipunculid worm. *Phascolosoma agassizii*. In: Wolfe DA (ed) Fate and effects of petroleum hydrocarbons in marine ecosystems and organisms. Pergamon Press, Elmsford, NY, pp 275–285

Anderson TD, Lydy MJ (2002) Increased toxicity to invertebrates associated with a mixture of atrazine and organophosphate insecticides. Environ Toxicol Chem 21:1507–1514

Anderson BS, Phillips BM, Hunt JW, Connor V, Richard N, Tjeerdema RS (2006) Identifying primary stressors impacting macroinvertebrates in the Salinas River (California, USA): relative effects of pesticides and suspended particles. Environ Pollut 141:402–408

ANZECC, ARMCANZ (2000) Australian and New Zealand guidelines for fresh and marine water quality. Report Australian and New Zealand Environment and Conservation Council and Agriculture and Resource Management Council of Australia and New Zealand, Canberra, Australia

Asfaw A, Ellersieck MR, Mayer FL (2003) Interspecies correlation estimations (ICE) for acute toxicity to aquatic organisms and wildlife. II. User manual and software. US Environmental Protection Agency Report No. EPA/600/R-03/106, Washington, DC, 20p + software

ASTM (1997) Standard test method for partition coefficient (n-octanol/water) estimation by liquid chromatography. Annual book of standards, E 1147-92. American Society for Testing and Materials, West Conshohocken, PA

ASTM (2001a) Practice for determination of hydrolysis rate constants of organic chemicals in aqueous solutions. Annual book of standards, E 895-89. American Society for Testing and Materials, West Conshohocken, PA

ASTM (2001b) Test method for determining a sorption constant (Koc) for an organic chemical in soil and sediments. Annual book of standards, E 1195-01. American Society for Testing and Materials, West Conshohocken, PA

ASTM (2002a) Guide for conducting bioconcentration tests with fishes and saltwater bivalve mollusks. Annual Book of Standards, E 1022-94. American Society for Testing and Materials, West Conshohocken, PA

ASTM (2002b) Test method for measurements of aqueous solubility. Annual book of standards, E 1148-02. American Society for Testing and Materials, West Conshohocken, PA

Bailey HC, Miller JL, Miller MJ, Wiborg LC, Deanovic L, Shed T (1997) Joint acute toxicity of diazinon and chlorpyrifos to *Ceriodaphnia dubia*. Environ Toxicol Chem 16:2304–2308

Barata C, Solayan A, Porte C (2004) Role of B-esterases in assessing toxicity of organophosphorus (chlorpyrifos, malathion) and carbamate (carbofuran) pesticides to *Daphnia magna*. Aquat Toxicol 66:125–139

Barry MJ, Logan DC, Vandam RA, Ahokas JT, Holdway DA (1995a) Effect of age and weight-specific respiration rate on toxicity of esfenvalerate pulse-exposure to the Australian crimson-spotted rainbow fish (*Melanotaenia fluviatilis*). Aquat Toxicol 32:115–126

Barry MJ, Ohalloran K, Logan DC, Ahokas JT, Holdway DA (1995b) Sublethal effects of esfenvalerate pulse-exposure on spawning and non-spawning Australian crimson-spotted rainbowfish (*Melanotaenia fluviatilis*). Arch Environ Contam Toxicol 28:459–463

Belden JB, Lydy MJ (2000) Impact of atrazine on organophosphate insecticide toxicity. Environ Toxicol Chem 19:2266–2274

Belden JB, Lydy MJ (2006) Joint toxicity of chlorpyrifos and esfenvalerate to fathead minnows and midge larvae. Environ Toxicol Chem 25:623–629

Berglund O, Larsson P, Ewald G, Okla L (2000) Bioaccumulation and differential partitioning of polychlorinated biphenyls in freshwater, planktonic food webs. Can J Fish Aquat Sci 57: 1160–1168

Borthwick PW, Clark JR, Montgomery RM, Patrick JM Jr, Lores EM (1985) Field confirmation of a laboratory-derived hazard assessment of the acute toxicity of fenthion to pink shrimp, *Penaeus duorarum*. In: Bahner RC, Hansen DJ (eds) Aquatic toxicology and hazard assessment: eighth symposium. ASTM STP 891, American Society of Testing and Materials, Philadelphia, PA, pp 177–189

Bowman J (1988) Acute flow through toxicity of chlorpyrifos to bluegill sunfish (*Lepomis macrochirus*): project ID 37189. Unpublished study prepared by Analytical Biochemistry Laboratories, Inc, 174p, MRID 40840904

Brannon JM, Pennington JC, Davis WM, Hayes C (1995) Fluoranthene K-DOC in sediment pore waters. Chemosphere 30:419–428

Brown R, Hugo J, Miller J, Harrington C (1997) Chlorpyrifos acute toxicity to the amphipod *Hyalella azteca*. Lab project No. 971095: 91/414 ANNEX I 8.3.4. Unpublished study prepared by the Dow Chemical Co., 27p, MRID 44345601

Brown MD, Carter J, Thomas D, Purdie DM, Kay BH (2002) Pulse-exposure effects of selected insecticides to juvenile Australian crimson-spotted rainbowfish (*Melanotaenia duboulayi*). J Econ Entomol 95:294–298

Bruce RD, Versteeg DJ (1992) A statistical procedure for modeling continuous toxicity data. Environ Toxicol Chem 11:1485–1494

Burgess D (1988) Acute flow through toxicity of chlorpyrifos to *Daphnia magna*: final Report No. 37190. Unpublished study prepared by Analytical Biochemistry Laboratories, Inc., 158p, MRID 40840902

Burgess RM, Pelletier MC, Gundersen JL, Perron MM, Ryba SA (2005) Effects of different forms of organic carbon on the partitioning and bioavailability of 4-nonylphenol. Environ Toxicol Chem 24:1609–1617

Burns LA (2004) Exposure analysis modeling system (EXAMS): user manual and system documentation, Revision G. Report US Environmental Protection Agency, Washington, DC

Burr IW (1942) Cumulative frequency functions. Ann Math Stat 13:215–232

Cairns J (1990) Lack of theoretical basis for predicting rate and pathways of recovery. Environ Manage 14:517–526

Cairns JJ, Dickson KL (1977) Recovery of streams from spills of hazardous materials. In: Cairns JJ, Dickson KL, Herricks EE (eds) Recovery and restoration of damaged ecosystems. University Press of Virginia, Charlottesville, VA, pp 24–42

Calamari D, Galassi S, Setti F, Vighi M (1983) Toxicity of selected chlorobenzenes to aquatic organisms. Chemosphere 12:253–262

Call DJ, Brooke LT, Knuth ML, Poirier SH, Hoglund MD (1985) Fish subchronic toxicity prediction model for industrial organic chemicals that produce narcosis. Environ Toxicol Chem 4:335–341

Campbell E, Palmer MJ, Shao Q, Warne M, Wilson D (2000) BurrliOZ: a computer program for calculating toxicant trigger values for the ANZECC and ARMCANZ water quality guidelines. In: National Water Quality Management Strategy, Australian and New Zealand Guidelines for Fresh and Marine Water Quality. Australian and New Zealand Environment and Conservation Council and Agricultural and Resource Management Council of Australia and New Zealand, Canberra, Australia. Available at http://www.cmis.csiro.au/Envir/burrlioz/

CCME (1999) A protocol for the derivation of water quality guidelines for the protection of aquatic life. Canadian environmental quality guidelines. Canadian Council of Ministers of the Environment, Ottawa

CDFG (1992a) Test No. 61, chronic, chlorpyrifos, *Ceriodaphnia dubia*. California Department of Fish and Game, Elk Grove, CA

CDFG (1992b) Test No. 133, acute, chlorpyrifos, *Neomysis mercedis*. California Department of Fish and Game, Elk Grove, CA

CDFG (1992c) Test No. 137, acute, chlorpyrifos, *Ceriodaphnia dubia*. California Department of Fish and Game, Elk Grove, CA

CDFG (1992d) Test No. 139, acute, chlorpyrifos, *Ceriodaphnia dubia*. California Department of Fish and Game, Elk Grove, CA

CDFG (1992e) Test No. 142, acute, chlorpyrifos, *Neomysis mercedis*. California Department of Fish and Game, Elk Grove, CA

CDFG (1992f) Test No. 143, acute, chlorpyrifos, *Neomysis mercedis*. California Department of Fish and Game, Elk Grove, CA

CDFG (1992 g) Test No. 150, acute, chlorpyrifos, *Ceriodaphnia dubia*. California Department of Fish and Game, Elk Grove, CA

CDFG (1992 h) Test No. 157, acute, diazinon, *Ceriodaphnia dubia*. California Department of Fish and Game, Elk Grove, CA

CDFG (1992i) Test No. 162, acute, diazinon, *Neomysis mercedis*. California Department of Fish and Game, Elk Grove, CA

CDFG (1992j) Test No. 163, acute, diazinon, *Ceriodaphnia dubia*. California Department of Fish and Game, Elk Grove, CA

CDFG (1992 k) Test No. 168, acute, diazinon, *Neomysis mercedis*. California Department of Fish and Game, Elk Grove, CA

CDFG (1998a) Test No. 122, acute, diazinon, *Ceriodaphnia dubia*. California Department of Fish and Game, Elk Grove, CA

CDFG (1998b) Test No. 132, acute, diazinon, *Physa* spp. California Department of Fish and Game, Elk Grove, CA

CDFG (1999) Test No. 61, 7-day chronic, chlorpyrifos, *Ceriodaphnia dubia*. California Department of Fish and Game, Elk Grove, CA

CDFG (2006a) State and federally listed endangered, threatened animals of California. California Natural Diversity Database. Available at www.dfg.ca.gov/hcpb/species/t_e_spp/tespp.shtml. California Department of Fish and Game, Sacramento, CA

CDFG (2006b) State and federally listed endangered, threatened, and rare plants of California. California Natural Diversity Database. Available at www.dfg.ca.gov/hcpb/species/t_e_spp/tespp.shtml. California Department of Fish and Game, Sacramento, CA

CDPR (2005) Registration desk manual. California Department of Pesticides Regulation, Sacramento, CA

Chapman PM, Fairbrother A, Brown D (1998) A critical evaluation of safety (uncertainty) factors for ecological risk assessment. Environ Toxicol Chem 17:99–108

Charles JR (1958) Final report on population manipulation studies in three Kentucky streams. Proc Southeast Assoc Game Fish Commrs 11:155–185

Chin YP, Gschwend PM (1992) Partitioning of poly cyclic aromatic hydrocarbons to marine porewater organic colloids. Environ Sci Technol 26:1621–1626

Chiou CT, Malcolm RL, Brinton TI, Kile DE (1986) Water solubility enhancement of some organic pollutants and pesticides by dissolved humic and fulvic-acids. Environ Sci Technol 20:502–508

Chiou CT, Kile DE, Brinton TI, Malcolm RL, Leenheer JA, Maccarthy P (1987) A comparison of water solubility enhancements of organic solutes by aquatic humic materials and commercial humic acids. Environ Sci Technol 21:1231–1234

CIBA-GEIGY (1987) Static acute toxicity of diazinon AG500 to bluegill (*Lepomis macrochirus*), EPA guidelines no. 72-1. Report for study conducted by Springborn Life Sciences, Inc., Wareham, MA for CIBA-GEIGY Corporation, Greensboro, NC

Cold A, Forbes VE (2004) Consequences of a short pulse of pesticide exposure for survival and reproduction of *Gammarus pulex*. Aquat Toxicol 67:287–299

Cook SF, Moore RL (1969) The effects of rotenone treatment on the insect fauna of a California stream. Trans Am Fish Soc 98:539–544

Corbet PS (1958) Some effects of DDT on the fauna of the Victoria Nile. Rev Zool Bot Afr 57: 73–95

Cornelissen G, Breedveld GD, Naes K, Oen AMP, Ruus A (2006) Bioaccumulation of native polycyclic aromatic hydrocarbons from sediment by a polychaete and a gastropod: freely dissolved concentrations and activated carbon amendment. Environ Toxicol Chem 25:2349–2355

Crane M (1997) Research needs for predictive multispecies tests in aquatic toxicology. Hydrobiologia 346:149–155

Crane M, Attwood C, Sheahan D, Morris S (1999) Toxicity and bioavailability of the organophosphorus insecticide pirimiphos methyl to the freshwater amphipod *Gammarus pulex* L. in laboratory and mesocosm systems. Environ Toxicol Chem 18:1456–1461

Crosby DG (1998) Environmental toxicology and chemistry. Oxford University Press, New York, NY

CSIRO (2001) BurrliOZ v. 1.0.13. Commonwealth Scientific and Industrial Research Organization, Australia

CVRWQCB (2004) The water quality control plan (basin plan) for the California Regional Water Quality Control Board Central Valley Region, 4th edn. the Sacramento and San Joaquin River basins. Central Valley Regional Water Quality Control Board, Rancho Cordova, CA

Day KE (1991) Effects of dissolved organic-carbon on accumulation and acute toxicity of fenvalerate, deltamethrin and cyhalothrin to *Daphnia magna* (Straus). Environ Toxicol Chem 10:91–101

Debruijn J, Busser F, Seinen W, Hermens J (1989) Determination of octanol-water partition coefficients for hydrophobic organic chemicals with the slow-stirring method. Environ Toxicol Chem 8:499–512

Delle Site A (2001) Factors affecting sorption of organic compounds in natural sorbent/water systems and sorption coefficients for selected pollutants. A review. J Phys Chem Ref Data 30:187–439

Denton D, Norberg-King TJ (1996) Whole effluent toxicity statistics: a regulatory perspective. In: Grothe DR, Dickson KL, Reed-Judkins DK (eds) Whole effluent toxicity testing: an evaluation of methods and prediction of receiving system impacts. SETAC Press, Pensacola, FL, pp 83–102

Denton DL, Fox JF, Fulk FA (2003) Enhancing toxicity performance by using a statistical criterion. Environ Toxicol Chem 22:2323–2328

Dermott RM, Spence HJ (1984) Changes in populations and drift of stream invertebrates following lampricide treatment. Can J Fish Aquat Sci 41:1695–1701

Dewolf W, Canton JH, Deneer JW, Wegman RCC, Hermens JLM (1988) Quantitative structure activity relationships and mixture-toxicity studies of alcohols and chlorohydrocarbons – reproducibility of effects on growth and reproduction of *Daphnia magna*. Aquat Toxicol 12:39–49

DiToro DM, Zarba CS, Hansen DJ, Berry WJ, Swartz RC, Cowan CE, Pavlou SP, Allen HE, Thomas NA, Paquin PR (1991) Technical basis for establishing sediment quality criteria for nonionic organic-chemicals using equilibrium partitioning. Environ Toxicol Chem 10:1541–1583

Eadie BJ, Morehead NR, Landrum PF (1990) 3-Phase partitioning of hydrophobic organic-compounds in Great Lakes waters. Chemosphere 20:161–178

ECB (2003) Technical guidance document on risk assessment in support of commission directive 93/67/EEC on risk assessment of new notified substances, commission regulation (EC) no. 1488/94 on risk assessment for existing substances, directive 98/8/EC of the European Parliament and of the Council concerning the placing of biocidal products on the market. Part II. Environmental risk assessment. European Chemicals Bureau Office for Publications of the European Communities, Luxembourg

ECETOC (1993) Technical Report No. 56 – aquatic toxicity data evaluation. European Centre for Ecotoxicology and Toxicology of Chemicals, Brussels. Available at http://www.ecetoc.org/technical-reports

ECOTOX (2006) ECOTOX code list. Report US Environmental Protection Agency, Washington, DC

Eidt DC (1981) Recovery of aquatic arthropod populations in a woodland stream after depletion – by fenitrothion treatment. Can Entomol 113:303–313

Ellersieck MR, Asfaw A, Mayer FL, Krause GF, Sun K, Lee G (2003) Acute-to-chronic estimation (ACE v. 2.0) with time-concentration-effect models: user manual and software. US Environmental Protection Agency Report No. EPA/600/R-03/107, Washington, DC, 26p, + software

El-Merhibi A, Kumar A, Smeaton T (2004) Role of piperonyl butoxide in the toxicity of chlorpyrifos to *Ceriodaphnia dubia* and *Xenopus laevis*. Ecotoxicol Environ Saf 57:202–212

Elson PF (1967) Effects on wild young salmon of spraying DDT over New Brunswick forests. J Fish Res Board Can 24:731–767

Emans HJB, Vanderplassche EJ, Canton JH, Okkerman PC, Sparenburg PM (1993) Validation of some extrapolation methods used for effect assessment. Environ Toxicol Chem 12:2139–2154

Erickson RJ, Stephan CE (1988) Calculation of the final acute value for water quality criteria for aquatic organisms. Report Environmental Research Laboratory-Duluth, US Environmental Protection Agency, Duluth, MN

Evans M, Hastings N, Peacock B (2000) Statistical distributions, 3rd edn. Wiley, New York, NY

Evers EHG, Smedes F (1993) Adsorptiegedrag van extreme hydrofobe verbindingen: PCDs, PAK's en dioxins. Bepalingsmethoden vertroebelen sorptiecoefficienten. Symposiumverslag Kontaminanten in Bodems en Sediment, Sorptie en Biologische Beschikbaarheid

EVS (1999) A critique of the ANZECC and ARMCANZ (1999) Water quality guidelines. Prepared for Minerals Council of Australia and Kwinana Industries Council. Final report. Vancouver, BC

Felsot AS (2005) A critical analysis of the draft report, "Amendments to the water quality control plan for the Sacramento River and San Joaquin River basins for the control of diazinon and chlorpyrifos runoff into the lower San Joaquin River," (Karkoski et al. 2004) and supporting documents. Prepared for the Central Valley Regional Water Quality Control Board, Sacramento, CA

Ferrari A, Venturino A, de D'Angelo AMP (2004) Time course of brain cholinesterase inhibition and recovery following acute and subacute azinphosmethyl, parathion and carbaryl exposure in the goldfish (*Carassius auratus*). Ecotoxicol Environ Saf 57:420–425

Finney DJ (1942) The analysis of toxicity tests on mixtures of poisons. Ann Appl Biol 29:82–94

Forbes VE, Calow P (1999) Is the per capita rate of increase a good measure of population-level effects in ecotoxicology? Environ Toxicol Chem 18:1544–1556

Forbes VE, Calow P (2002) Species sensitivity distributions revisited: a critical appraisal. Hum Ecol Risk Assess 8:473–492

Forbes VE, Cold A (2005) Effects of the pyrethroid esfenvalerate on life-cycle traits and population dynamics of *Chironomus riparius* – importance of exposure scenario. Environ Toxicol Chem 24:78–86

Fox DR (1999) Setting water quality guidelines – a statistician's perspective. SETAC News 19:17–18

Fredeen FJH (1975) Effects of a single injection of methoxyclor black-fly larvicide on insect larvae in a 161-km (100-Mile) section of North Saskatchewan River. Can Entomol 107:807–817

Fredeen FJH (1983) Trends in numbers of aquatic invertebrates in a large Canadian river during four years of black fly larviciding with methoxychlor (Diptera: Simuliidae). Quaestiones Entomologicae 19:53–92

Galassi S, Mingazzini M, Vigano L, Cesareo D, Tosato ML (1988) Approaches to modeling toxic responses of aquatic organisms to aromatic hydrocarbons. Ecotoxicol Environ Saf 16:158–169

Garbarini DR, Lion LW (1986) Influence of the nature of soil organics on the sorption of toluene and trichloroethylene. Environ Sci Technol 20:1263–1269

Gauthier TD, Seitz WR, Grant CL (1987) Effects of structural and compositional variations of dissolved humic materials on pyrene Koc values. Environ Sci Technol 21:243–248

Geiger DL, Call DJ, Brooke LT (1988) Acute toxicities of organic chemicals to fathead minnows (*Pimephales promelas*), Volume IV. Center for Lake Superior Environmental Studies, University of Wisconsin-Superior, Superior, WI, pp 195–197

Ghetti PF, Gorbi G (1985) Effects of acute pollution on macroinvertebrates in a stream. Verhandlungen IVL 22:2426–2431

Giesy JP, Solomon KR, Coats JR, Dixon KR, Giddings JM, Kenaga EE (1999) Chlorpyrifos: ecological risk assessment in North American aquatic environments. Rev Environ Contam Toxicol 160:1–129

GLEC (2003) Draft compilation of existing guidance for the development of site-specific water quality objectives in the state of California. Great Lakes Environmental Center, Columbus, OH

Grothe DR, Kickson KL, Reed-Judkins DK (eds) (1996) Whole effluent toxicity testing: an evaluation of methods and prediction of receiving system impacts. SETAC Press, Pensacola, FL

Gustafson KE, Dickhut RM (1997) Distribution of polycyclic aromatic hydrocarbons in southern Chesapeake Bay surface water: evaluation of three methods for determining freely dissolved water concentrations. Environ Toxicol Chem 16:452–461

Halter MT, Johnson HE (1977) A model system to study the desorption and biological availability of PCB in hydrosoils. In: Mayer FL, Hamelink JL (eds) Aquatic Toxicity and Hazard Evaluation ASTM STP 634. American Society for Testing and Materials, Philadelphia, PA, pp 178–195

Hanson ML, Sanderson H, Solomon KR (2003) Variation, replication, and power analysis of *Myriophyllum* spp. microcosm toxicity data. Environ Toxicol Chem 22:1318–1329

Harmon SM, Specht WL, Chandler GT (2003) A comparison of the daphnids *Ceriodaphnia dubia* and *Daphnia ambigua* for their utilization in routine toxicity testing in the Southeastern United States. Arch Environ Contam Toxicol 45:79–85

Harrison AD, Rattray EA (1966) Biological effects of mollusciciding natural waters. S Afr J Sci 62:236–241

Hastings E, Kittams WH, Pepper JH (1961) Repopulation by aquatic insects in streams sprayed with DDT. Ann Entomol Soc Am 54:436–437

Heckmann LH, Friberg N (2005) Macroinvertebrate community response to pulse exposure with the insecticide lambda-cyhalothrin using in-stream mesocosms. Environ Toxicol Chem 24: 582–590

Hermens J, Canton H, Janssen P, Dejong R (1984) Quantitative structure activity relationships and toxicity studies of mixtures of chemicals with anesthetic potency – acute lethal and sublethal toxicity to *Daphnia magna*. Aquat Toxicol 5:143–154

Hoekstra JA, Van Ewijk (1993) Alternatives for the no-observed-effect level. Environ Toxicol Chem 12:187–194

Hoffman CH, Drooz AT (1953) Effects of a C-47 airplane application of DDT on fish-food organisms in two Pennsylvania watersheds. Am Midl Nat 50: 172–188

Holcombe GW, Phipps GL, Tanner DK (1982) The acute toxicity of Kelthane, Dursban, Disulfoton, Pydrin, and Permethrin to Fathead Minnows *Pimephales promelas* and Rainbow Trout *Salmo gairdneri*. Environ Pollut Ser A Ecol Biol 29:167–178

Holdway DA, Barry MJ, Logan DC, Robertson D, Young V, Ahokas JT (1994) Toxicity of pulse-exposed fenvalerate and esfenvalerate to larval Australian crimson-spotted rainbow fish (*Melanotaenia fluviatilis*). Aquat Toxicol 28:169–187

Hose GC, Van Den Brink PJ (2004) Confirming the species-sensitivity distribution concept for endosulfan using laboratory, mesocosm, and field data. Arch Environ Contam Toxicol 47: 511–520

Host GE, Regal RR, Stephan CE (1995) Analyses of acute and chronic data for aquatic life. US Environmental Protection Agency, Washington, DC

Howard PH (1991) Handbook of environmental fate and exposure data for organic chemicals. Pesticides, vol III. CRC Press, Boca Raton, FL

Huckins JN, Tubergen MW, Manuweera GK (1990) Semipermeable membrane devices containing model lipid: a new approach to monitoring the bioavailability of lipophilic contaminants and estimating their bioconcentration potential. Chemosphere 20:533–552

Huckins JN, Petty JD, Lebo JA, Almeida FV, Booij K, Alvarez DA, Clark RC, Mogensen BB (2002) Development of the permeability/performance reference compound approach for in situ calibration of semipermeable membrane devices. Environ Sci Technol 36:85–91

Huckins JN, Prest HF, Petty JD, Lebo JA, Hodgins MM, Clark RC, Alvarez DA, Gala WR, Steen A, Gale R, Ingersoll CI, Steen A (2004) Overview and comparison of lipid-containing semipermeable membrane devices and oysters (*Crassostrea gigas*) for assessing organic chemical exposure. Environ Toxicol Chem 23:1617–1628

Hyder AH, Overmyer JP, Noblet R (2004) Influence of developmental stage on susceptibilities and sensitivities of *Simulium vittatum* IS-7 and *Simulium vittatum* IIIL-1 (Diptera: Simuliidae) to chlorpyrifos. Environ Toxicol Chem 23:2856–2862

Irmer U, Markard C, Blondzik K, Gottschalk C, Kussatz C, Rechenberg B, Schudoma D (1995) Quality targets for concentrations of hazardous substances in surface waters in Germany. Ecotoxicol Environ Saf 32:233–243

Jacobi GZ, Degan DJ (1977) Aquatic macroinvertebrates in a small Wisconsin trout stream before, during and two years after treatment with the fish toxicant antimycin. US Bureau of Sport Fisheries and Wildlife. Investig Fish Control 81:1–24

Jarvinen AW, Tanner DK (1982) Toxicity of selected controlled release and corresponding unformulated technical grade pesticides to the Fathead Minnow *Pimephales promelas*. Environ Pollut Series a-Ecol Biol 27:179–195

Jarvinen AW, Tanner DK, Kline ER (1988) Toxicity of chlorpyrifos, endrin, or fenvalerate to fathead minnows following episodic or continuous exposure. Ecotoxicol Environ Saf 15: 78–95

Jeffrey KA, Beamish FWH, Ferguson SC, Kolton RJ, Macmahon PD (1986) Effects of the lampricide, 3-trifluoromethyl-4-nitrophenol (TFM) on the macroinvertebrates within the hyporheic region of a small stream. Hydrobiologia 134:43–51

JMP (2004) Statistical discovery software. Version 5.1.2. SAS Institute, Inc., Cary, NC

Keenleyside MHA (1959) Effects of spruce budworm control on salmon and other fishes in New Brunswick. Can Fish Cult 24:17–22

Kenaga EE (1982) Predictability of chronic toxicity from acute toxicity of chemicals in fish and aquatic invertebrates. Environ Toxicol Chem 1:347–358

Kersting K, Van Wijngaarden R (1992) Effects of chlorpyrifos on a microecosystem. Environ Toxicol Chem 11:365–372

Konemann H (1981) Quantitative structure-activity-relationships in fish toxicity studies. 1. Relationship for 50 industrial pollutants. Toxicology 19:209–221

Kooijman S (1987) A safety factor for LC50 values allowing for differences in sensitivity among species. Water Res 21:269–276

Kraufvelin P (1999) Baltic hard bottom mesocosms unplugged: replicability, repeatability and ecological realism examined by non-parametric multivariate techniques. J Exp Mar Biol Ecol 240:229–258

Kuhn R, Pattard M, Pernak KD, Winter A (1989) Results of the harmful effects of water pollutants to *Daphnia magna* in the 21-day reproduction test. Water Res 23:501–510

Kukkonen J, Oikari A (1991) Bioavailability of organic pollutants in boreal waters with varying levels of dissolved organic material. Water Res 25:455–463

Lange R, Hutchinson TH, Croudace CP, Siegmund F (2001) Effects of the synthetic estrogen 17 alpha-ethinylestradiol on the life-cycle of the fathead minnow (*Pimephales promelas*). Environ Toxicol Chem 20:1216–1227

Laor Y, Farmer WJ, Aochi Y, Strom PF (1998) Phenanthrene binding and sorption to dissolved and to mineral-associated humic acid. Water Res 32:1923–1931

Lauridsen RB, Friberg N (2005) Stream macroinvertebrate drift response to pulsed exposure of the synthetic pyrethroid lambda-cyhalothrin. Environ Toxicol 20:513–521

Lee GH, Ellersieck MR, Mayer FL, Krause GF (1995) Predicting chronic lethality of chemicals to fishes from acute toxicity test data – multifactor probit analysis. Environ Toxicol Chem 14:345–349

Lepper P (2000) Towards the derivation of quality standards for priority substances in the context of the Water Framework Directive. Final report of the study contract no. B4-3040/2000/30673/MAR/E1. Fraunhofer-Institute Molecular Biology and Applied Ecology, Munich

Liess M, Schulz R (1999) Linking insecticide contamination and population response in an agricultural stream. Environ Toxicol Chem 18:1948–1955

Lillebo HP, Shaner S, Carlson D, Richard N (1988) Water quality criteria for selenium and other trace elements for protection of aquatic life and its uses in the San Joaquin Valley. Technical committee report: regulation of agricultural drainage to the San Joaquin River. Appendix D. California State Water Resources Control Board, Sacramento, CA

Little JD (1966) Reclamation of Pine Creek, Tennessee. Proc Southeast Assoc Game Fish Commrs 19:302–315

Liu WP, Gan JJ, Lee S, Kabashima JN (2004) Phase distribution of synthetic pyrethroids in runoff and stream water. Environ Toxicol Chem 23:7–11

Lu Y, Wang Z (2003) Accumulation of organochlorinated pesticides by triolein-containing semipermeable membrane device (triolein-SPMD) and rainbow trout. Water Res 37:2419–2425

Lydy MJ, Belden JB, Wheelock CE, Hammock BD, Denton DL (2004) Challenges in regulating pesticide mixtures. Ecol Soc 9:1

Lydy MJ, Austin KR (2005) Toxicity assessment of pesticide mixtures typical of the Sacramento-San Joaquin Delta using *Chironomus tentans*. Arch Environ Contam Toxicol 48:49–55

Mackay D (1982) Correlation of bioconcentration factors. Environ Sci Technol 16:274–278

Mackay D, Shiu WY, Ma KC (1997) Illustrated handbook of physical-chemical properties and environmental fate for organic chemicals. CRC Press, Boca Raton, FL

Mackay D (2001) Multimedia environmental fate models: the fugacity approach, 2nd edn. Lewis Publishers, Boca Raton, FL

Maltby L, Blake N, Brock TCM, Van Den Brink PJ (2005) Insecticide species sensitivity distributions: importance of test species selection and relevance to aquatic ecosystems. Environ Toxicol Chem 24:379–388

Maltby L, Brock TCM, Van Den Brink PJ (2009) Fungicide risk assessment for aquatic ecosystems: importance of interspecific variation, toxic mode of action, and exposure regime. Environ Sci Technol 43:7556–7563

Maund SJ, Hamer MJ, Lane MCG, Farrelly E, Rapley JH, Goggin UM, Gentle WE (2002) Partitioning, bioavailability, and toxicity of the pyrethroid insecticide cypermethrin in sediments. Environ Toxicol Chem 21:9–15

Mayer FL, Krause GF, Buckler DR, Ellersieck MR, Lee GH (1994) Predicting chronic lethality of chemicals to fishes from acute toxicity test data – concepts and linear-regression analysis. Environ Toxicol Chem 13:671–678

Mayer FL, Ellersieck MR, Krause GF, Sun K, Lee G, Buckler DR (2002) Time-concentration-effect models in predicting chronic toxicity from acute toxicity data. In: Crane M, Newman MC, Chapman PF, Fenlon J (eds) Risk assessment with time to event models. Lewis Publishers, Boca Raton, FL

McCarthy JF, Jimenez BD, Barbee T (1985) Effect of dissolved humic material on accumulation of polycyclic aromatic-hydrocarbons – structure activity relationships. Aquat Toxicol 7:15–24

Meehan WR, Sheridan WL (1966) Effects of toxaphene on fishes and bottom fauna of Big Kitoi Creek, Afognak Island, Alaska. Fish and Wildlife Service Resource Publication No. 12. US Department of the Interior, Washington, DC

Menconi M, Beckman J (1996) Hazard assessment of the insecticide methomyl to aquatic organisms in the San Joaquin River system. Administrative Report 96-6. California Department of Fish and Game, Environmental Service Division, Rancho Cordova, CA

MHSPE (1994) Intervention values and target values – soil quality standards. Ministry of Housing, Spatial Planning and Environment. Directorate-General for Environmental Protection, The Hague, The Netherlands

Minckley WL, Mihalick P (1981) Effects of chemical treatment for fish eradication on stream-dwelling invertebrates. J Arizona-Nevada Acad Sci 16:79–82

MITI (1992) Biodegradation and bioaccumulation data of existing chemicals based on the CSCL Japan. Japan Chemical Industry Ecology-Toxicology & Information Center. Ministry of International Trade and Industry, Basic Industries Bureau, Chemical Products Safety Division

Moore DRJ, Caux PY (1997) Estimating low toxic effects. Environ Toxicol Chem 16: 794–801

Morrison BRS (1977) The effects of rotenone on the invertebrate fauna of three hill streams in Scotland. Fisheries Manage 8:128–139

Moye WC, Luckmann WH (1964) Fluctuations in populations of certain aquatic insects following application of aldrin granules to Sugar Creek Iroquois County Illinois. J Econ Entomol 57: 318–322

Mu XY, LeBlanc GA (2004) Synergistic interaction of endocrine-disrupting chemicals: model development using an ecdysone receptor antagonist and a hormone synthesis inhibitor. Environ Toxicol Chem 23:1085–1091

Nabholz JV (1991) Environmental hazard and risk assessment under the United States Toxic Substances Control Act. Sci Total Environ 109:649–665

Naddy RB, Johnson KA, Klaine SJ (2000) Response of Daphnia magna to pulsed exposures of chlorpyrifos. Environ Toxicol Chem 19:423–431

Neff JM (1979) Polycyclic aromatic hydrocarbons in the aquatic environment. Applied Science, London

Niemi GJ, Devore P, Detenbeck N, Taylor D, Lima A, Pastor J, Yount JD, Naiman RJ (1990) Overview of case-studies on recovery of aquatic systems from disturbance. Environ Manage 14:571–587

Nikunen E, Leinonen R, Kemilainen B, Kultamaa A (2003) Environment guide 71 – environmental properties of chemicals. Finnish Environment Institute. Helsinki, Finland

North Carolina DENR (2003) Redbook: surface water and wetland standards. Division of Water Quality North Carolina Department of Environment and Natural Resources, Raleigh, NC

OECD (1981) Test No. 112: dissociation constants in water. OECD publishing. Available at http://browse.oecdbookshop.org/oecd/pdfs/browseit/9711201E.PDF

OECD (1995a) OECD environment monographs No. 92, OECD environmental health and safety publications, series on testing and assessment, No. 3, guidance document for aquatic effects assessment. Organization for Economic Co-operation and Development, Paris

OECD (1995b) Test No. 105: water solubility. OECD publishing. Available at http://www.oecd.org/dataoecd/17/13/1948185.pdf

OECD (1996) Test No. 305: bioconcentration: flow-through fish test. OECD publishing. Available at http://browse.oecdbookshop.org/oecd/pdfs/browseit/9730501E.PDF

OECD (2000) Test No. 106: adsorption – desorption using a batch equilibrium method. OECD publishing. Available at http://browse.oecdbookshop.org/oecd/pdfs/browseit/9710601E.PDF

OECD (2001) Test No. 121: estimation of the adsorption coefficient (Koc) on soil and on sewage sludge using high performance liquid chromatography (HPLC). OECD publishing. Available at http://browse.oecdbookshop.org/oecd/pdfs/browseit/9712101E.PDF

OECD (2004) Test No. 111: hydrolysis as a function of pH. OECD publishing. Available at http://browse.oecdbookshop.org/oecd/pdfs/browseit/9711101E.PDF

Okkerman PC, Van Den Plassche EJ, Slooff W, Van Leeuwen CJ, Canton JH (1991) Ecotoxicological effects assessment: a comparison of several extrapolation procedures. Ecotox Environ Saf 21:182–193

Okkerman PC, Vanderplassche EJ, Emans HJB, Canton JH (1993) Validation of some extrapolation methods with toxicity data derived from multiple species experiments. Ecotoxicol Environ Saf 25:341–359

Olmstead AW, LeBlanc GA (2005) Toxicity assessment of environmentally relevant pollutant mixtures using a heuristic model. Integr Environ Assess Manage 1:114–122

Onsager L (1927) On the theory of electrolytes. II. Physikalische Zeitschrift 28:277–298

PAN (2006) Pesticide Action Network Pesticide Database. Available at http://www.pesticideinfo.org/Index.html

PapeLindstrom PA, Lydy MJ (1997) Synergistic toxicity of atrazine and organophosphate insecticides contravenes the response addition mixture model. Environ Toxicol Chem 16:2415–2420

Parsons JT, Surgeoner GA (1991a) Acute toxicities of permethrin, fenitrothion, carbaryl and carbofuran to mosquito larvae during single-pulse or multiple-pulse exposures. Environ Toxicol Chem 10:1229–1233

Parsons JT, Surgeoner GA (1991b) Effect of exposure time on the acute toxicities of permethrin, fenitrothion, carbaryl and carbofuran to mosquito larvae. Environ Toxicol Chem 10: 1219–1227

Persoone G, Janssen CR (1994) Field validation of predictions based on laboratory toxicity tests. In: Hill IR, Heimbach F, Leeuwangh P, Matthiessen P (eds) Freshwater field tests for hazard assessment of chemicals. CRC Press, Boca Raton, FL, pp 379–397

Peterson JL, Jepson PC, Jenkins JJ (2001) Effect of varying pesticide exposure duration and concentration on the toxicity of carbaryl to two field-collected stream invertebrates, *Calineuria californica* (Plecoptera: Perlidae) and *Cinygma* sp. (Ephemeroptera: Heptageniidae). Environ Toxicol Chem 20:2215–2223

Phipps GL, Holcombe GW (1985) A method for aquatic multiple species toxicant testing – acute toxicity of 10 chemicals to 5 vertebrates and 2 invertebrates. Environ Pollut Series a-Ecol Biol 38:141–157

Phillips TA, Wu JG, Summerfelt RC, Atchison GJ (2002) Acute toxicity and cholinesterase inhibition in larval and early juvenile walleye exposed to chlorpyrifos. Environ Toxicol Chem 21:1469–1474

Phillips TA, Summerfelt RC, Wu J, Laird DA (2003) Toxicity of chlorpyrifos adsorbed on humic colloids to larval walleye (*Stizostedion vitreum*). Arch Environ Contam Toxicol 45: 258–263

PHYSPROP (2006) Physical Properties Database. Available at www.syrres.com/esc/physprop.htm. SRC Inc

Plackett RL, Hewlett PS (1952) Quantal responses to mixtures of poisons. J R Stat Soc Series B Stat Methodol 14:141–163

Prest H, Petty JD, Huckins JN (1998) Validity of using semipermeable membrane devices for determining aqueous concentrations of freely dissolved PAHs. Environ Toxicol Chem 17: 535–536

Printes LB, Callaghan A (2004) A comparative study on the relationship between acetyl cholinesterase activity and acute toxicity in Daphnia magna exposed to anticholinesterase insecticides. Environ Toxicol Chem 23:1241–1247

Reynaldi S, Liess M (2005) Influence of duration of exposure to the pyrethroid fenvalerate on sublethal responses and recovery of Daphnia magna Straus. Environ Toxicol Chem 24: 1160–1164

Rider CV, LeBlanc GA (2005) An integrated addition and interaction model for assessing toxicity of chemical mixtures. Toxicol Sci 87:520–528

RIVM (2001) Guidance document on deriving environmental risk limits in The Netherlands, report 601501012. National Institute of Public Health and the Environment, Bilthoven, The Netherlands

Roesijadi G, Anderson JW, Blaylock JW (1978a) Uptake of hydrocarbons from marine sediments contaminated with Prudhoe Bay crude oil – influence of feeding type of test species and availability of polycyclic aromatic hydrocarbons. J Fish Res Board Can 35:608–614

Roesijadi G, Woodruff DL, Anderson JW (1978b) Bioavailability of naphthalenes from marine sediments artificially contaminated with Prudhoe Bay crude oil. Environ Pollut 15:223–229

Rogers HR (1993) Speciation and partitioning of priority organic contaminants in estuarine waters. Colloids Surf A 73:229–235

Rossi SS (1977) Bioavailability of petroleum hydrocarbons from water, sediments and detritus to the marine annelid, *Neanthes arenaceodentata*. American Petroleum Institute, Washington, DC, pp 621–626

Roux DJ, Jooste SHJ, MacKay HM (1996) Substance-specific water quality criteria for the protection of South African freshwater ecosystems: methods for derivation and initial results for some inorganic toxic substances. S Afr J Sci 92:198–206

Samsoe-Petersen L, Pedersen F (1995) Water quality criteria for selected priority substances, working report TI 44. Water Quality Institute, Danish Environmental Protection Agency, Copenhagen, Denmark

Sanderson H (2002) Pesticide studies – replication of micro/mesocosm studies. Environ Sci Pollut Res 6:429–435

Sangster Research Laboratories (2004) LOGKOW, a databank of evaluated octanol-water partition coefficients (Log P), http://logkow.cisti.nrc.ca/logkow/intro.html. Canadian National Committee for CODATA, Montreal, QC

Schnürer Y, Persson M, Nilsson M, Nordgren A, Giesler R (2006) Effects of surface sorption on microbial degradation of glyphosate. Environ Sci Technol 40:4145–4150

Schulz R, Liess M (2000) Toxicity of fenvalerate to caddisfly larvae: chronic effects of 1-vs 10-hr pulse-exposure with constant doses. Chemosphere 41:1511–1517

Schulz R, Liess M (2001) Toxicity of aqueous-phase and suspended particle-associated fenvalerate: chronic effects after pulse-dosed exposure of *Limnephilus lunatus* (Trichoptera). Environ Toxicol Chem 20:185–190

Schwarzenbach RP, Gschwend PM, Imboden DM (1993) Environmental organic chemistry. Wiley, New York, NY

Segner H (2005) Developmental, reproductive, and demographic alterations in aquatic wildlife: establishing causality between exposure to endocrine-active compounds (EACs) and effects. Acta Hydrochim Hydrobiol 33:17–26

Shen L, Wania F (2005) Compilation, evaluation, and selection of physical-chemical property data for organochlorine pesticides. J Chem Eng Data 50:742–768

Siepmann S, Jones MR (1998) Hazard assessment of the insecticide carbaryl to aquatic organisms in the Sacramento-San Joaquin river system, Administrative Report 98-1. California Department of Fish and Game, Office of Spill Prevention and Response, Rancho Cordova, CA

Siepmann S, Finlayson B (2000) Water quality criteria for diazinon and chlorpyrifos, Administrative Report 00-3. California Department of Fish and Game, Rancho Cordova, CA

Solomon KR, Giddings JM, Maund SJ (2001) Probabilistic risk assessment of cotton pyrethroids: I. Distributional analyses of laboratory aquatic toxicity data. Environ Toxicol Chem 20:652–659

Solomon KR, Takacs P (2002) Probabilistic risk assessment using species sensitivity distributions. In: Posthuma L, Suter GW, Traas TP (eds) Species sensitivity distributions in ecotoxicology. Lewis Publishers, New York, NY, pp 285–314

Staples CA, Dickson KL, Rodgers JHJ, Saleh FY (1985) A model for predicting the influence of suspended sediments on the bioavailability of neutral organic chemicals in the water compartment. In: Cardwell RD, Purdy R, Bahner RC (eds) Aquatic toxicology and hazard assessment: seventh symposium ASTM STP 854. American Society for Testing and Materials, Philadelphia, PA, pp 417–428

Steinberg CEW, Sturm A, Kelbel J, Lee SK, Hertkorn N, Freitag D, Kettrup AA (1992) Changes of acute toxicity of organic chemicals to *Daphnia magna* in the presence of dissolved humic material (DHM). Acta Hydrochim Hydrobiol 20:326–332

Steinberg CEW, Xu Y, Lee SK, Freitag D, Kettrup A (1993) Effect of dissolved humic material (DHM) on bioavailability of some organic xenobiotics to *Daphnia magna*. Chem Spec Bioavail 5:1–9

Stephan CE (1985) Are the "Guidelines for deriving numerical national water quality criteria for the protection of aquatic life and its uses" based on sound judgments? In: Cardwell RD, Purdy R, Bahner RC (eds) Aquatic toxicology and hazard assessment: seventh symposium, ASTM STP 854. American Society for Testing and Materials, Philadelphia, PA, pp 515–526

Stephan CE, Rogers JW (1985) Advantages of using regression analysis to calculate results of chronic toxicity tests. Aquatic Toxicology and Hazard Assessment: eighth symposium. American Society for Testing and Materials, pp 328–338

Sun K, Krause GF, Mayer FL, Ellersieck MR, Basu AP (1995) Predicting chronic lethality of chemicals to fishes from acute toxicity test data – theory of accelerated life testing. Environ Toxicol Chem 14:1745–1752

Suter GWI, Barnthouse LW (1993) Assessment concepts. In: Suter GWI (ed) Ecological risk assessment. Lewis Publishers, Boca Raton, FL, pp 21–47

Takahashi N, Ito M, Mikami N, Matsuda T, Miyamoto J (1988) Identification of reactive oxygen species generated by irradiation of aqueous humic-acid solution. J Pestic Sci 13: 429–435

TenBrook PL, Tjeerdema RS, Hann P, Karkoski J (2009) Methods for deriving pesticide aquatic life criteria. Rev Environ Contamin Toxicol 199:19–109

Torblaa RL (1968) Effects of lamprey larvicides on invertebrates in streams. US Department of the Interior, Washington, DC

Traas TP, VandeMeent D, Posthuma L, Hamers T, Kater BJ, De Zwart D, Aldenberg T (2002) The potentially affected fraction as a measure of ecological risk. In: Posthuma L, Suter GWI, Traas TP (eds) Species sensitivity distributions in ecotoxicology. Lewis Publishers, New York, NY, pp 315–344

US Code Title 7 (1947) Federal Insecticide, Fungicide and Rodenticide Act. Code of Federal Regulations, Title 7. As amended through 2004 Ed

US Code Title 40 (2009) Terrestrial and aquatic non-target organisms data requirements table. Federal Insecticide, Fungicide and Rodenticide Act. Code of Federal Regulations, Title 40, Volume 23, Part 158.630. As amended through 2009 Ed

USEPA (1980a) Ambient water quality criteria for aldrin/dieldrin. US Environmental Protection Agency, Washington, DC

USEPA (1980b) Ambient water quality criteria for chlordane, EPA 440/5-80-027. US Environmental Protection Agency, Washington, DC

USEPA (1980c) Ambient water quality criteria for endosulfan, EPA 440/5-80-046. US Environmental Protection Agency, Washington, DC

USEPA (1980d) Ambient water quality criteria for endrin, EPA 440/5-80-047. US Environmental Protection Agency, Washington, DC

USEPA (1980e) Ambient Water Quality Criteria for Heptachlor, EPA 440/5-80-062. US Environmental Protection Agency, Washington, DC

USEPA (1980f) Ambient Water Quality Criteria for Hexachlorocyclohexane, EPA 440/5-80-054. US Environmental Protection Agency, Washington, DC

USEPA (1980 g) Ambient Water Quality Criteria for DDT, EPA 440/5-80-038. US Environmental Protection Agency, Washington, DC

USEPA (1985) Guidelines for deriving numerical national water quality criteria for the protection of aquatic organisms and their uses, PB-85-227049. US Environmental Protection Agency, National Technical Information Service, Springfield, VA

USEPA (1986a) Ambient water quality criteria for chlorpyrifos, EPA 440/5-86-005. US Environmental Protection Agency, Washington, DC

USEPA (1986b) Ambient Water Quality Criteria for Toxaphene, EPA 440/5-86-006. US Environmental Protection Agency, Washington, DC

USEPA (1991) Technical Support Document for Water Quality-based Toxics Control, EPA/505/2-90-001. US Environmental Protection Agency, Washington, DC

USEPA (1993) 40 CFR 158.490 Wildlife and aquatic organisms data requirements. Code of Federal Regulations. US Environmental Protection Agency, Washington, DC

USEPA (1994) Water quality standards handbook. EPA-823-B-94-005. US Environmental Protection Agency, Washington, DC
USEPA (1996a) Product properties test guidelines. OPPTS 830.7560. Partition coefficient (n-octanol/water), generator column method, EPA 712-C-96-039. US Environmental Protection Agency, Washington, DC
USEPA (1996b) Product properties test guidelines. OPPTS 830.7550. Partition coefficient (n-octanol/water), shake flask method, 712-C-96-038. US Environmental Protection Agency, Washington, DC
USEPA (2002a) Understanding and accounting for method variability in whole effluent toxicity application under the National Pollutant Discharge Elimination System Program. EPA/833R-00-003. US EPA Office of Water, Washington, DC
USEPA (2002b) Short-term methods for estimating the chronic toxicity of effluents and receiving waters to freshwater organisms, 4th Ed. EPA-821-R-02-013. US Environmental Protection Agency, Washington, DC
USEPA (2003a) Water quality guidance for the Great Lakes system. Fed Regist 40
USEPA (2003b) Ambient aquatic life water quality criteria for tributyltin (TBT) – Final, EPA 822-R-03-031. US Environmental Protection Agency
USEPA (2003c) Draft ambient life water quality criteria for atrazine, EPA-822-R-03-023. Office of Water, Health and Ecological Criteria Division, US Environmental Protection Agency, Washington, DC
USEPA (2003d) 2003 Draft update of ambient water quality criteria for copper, EPA 822-R-03-026. US Environmental Protection Agency, Washington, DC
USEPA (2005a) Aquatic life ambient water quality criteria, diazinon, final, EPA-822-R-05-006. US Environmental Protection Agency, Washington, DC
USEPA (2005b) Science Advisory Board consultation document, proposed revisions to aquatic life guidelines, water-based criteria. US Environmental Protection Agency, Washington, DC
USEPA (2006) Recognition and management of pesticide poisonings, 5th edn. Section IV, Chapter 16 Fumigants. US Environmental Protection Agency, Washington, DC
USGS (2000) The virtual fish: SPMD basics. US Geological Survey Columbia Environmental Research Center
Van den Berg M, Birnbaum L, Bosveld ATC, Brunstrom B, Cook P, Feeley M, Giesy JP, Hanberg A, Hasegawa R, Kennedy SW, Kubiak T, Larsen JC, Van Leeuwen FXR, Liem AKD, Nolt C, Peterson RE, Poellinger L, Safe S, Schrenk D, Tillitt D, Tysklind M, Younes M, Waern F, Zacharewski T (1998) Toxic equivalency factors (TEFs) for PCBs, PCDDs, PCDFs for humans and wildlife. Environ Health Perspect 106:775–792
Van Der Hoeven N, Gerritsen AAM (1997) Effects of chlorpyrifos on individuals and populations of *Daphnia pulex* in the laboratory and field. Environ Toxicol Chem 16: 2438–2447
Van Der Hoeven N, Noppert F, Leopold A (1997) How to measure no effect. 1. Towards a new measure of chronic toxicity in ecotoxicology, introduction and workshop results. Environmetrics 8:241–248
Van Leeuwen CJ, Adema DMM, Hermens J (1990) Quantitative structure-activity-relationships for fish early life stage toxicity. Aquat Toxicol 16:321–334
Van Leeuwen CJ, Vanderzandt PTJ, Aldenberg T, Verhaar HJM, Hermens JLM (1992) Application of QSARs, extrapolation and equilibrium partitioning in aquatic effects assessment. 1. Narcotic industrial pollutants. Environ Toxicol Chem 11:267–282
Van Straalen NM, Denneman CAJ (1989) Ecotoxicological evaluation of soil quality criteria. Ecotoxicol Environ Saf 18:241–251
Van Straalen NM, Van Leeuwen CJ (2002) European history of species sensitivity distributions. In: Posthuma L, Suter GWI, Traas TP (eds) Species sensitivity distributions in ecotoxicology. Lewis Publishers, New York, NY pp 19–34
Van Vlaardingen PLA, Traas TP, Wintersen AM, Aldenberg T (2004) ETX 2.0 A program to calculate hazardous concentrations and fraction affected, based on normally distributed

toxicity data. National Institute of Public Health and the Environment (RIVM), Bilthoven, The Netherlands

Van Wijngaarden R, Leeuwangh P, Lucassen WGH, Romijn K, Ronday R, Vandervelde R, Willigenburg W (1993) Acute toxicity of chlorpyrifos to fish, a newt, and aquatic invertebrates. Bull Environ Contam Toxicol 51:716–723

Varó I, Serrano R, Pitarch E, Amat F, Lopez FJ, Navarro JC (2002) Bioaccumulation of chlorpyrifos through an experimental food chain: study of protein HSP70 as biomarker of sublethal stress in fish. Arch Environ Contam Toxicol 42:229–235

Veith GD, Call DJ, Brooke LT (1983) Structure toxicity relationships for the fathead minnow, *Pimephales promelas* – narcotic industrial chemicals. Can J Fish Aquat Sci 40:743–748

Verhaar HJM, Van Leeuwen CJ, Bol J, Hermens JLM (1994) Application of QSARs in risk management of existing chemicals. SAR QSAR Environ Res 2:39–58

Verschueren K (2009) Handbook of environmental data on organic chemicals, 5th edn. Wiley, Hoboken, NJ

Versteeg DJ, Belanger SE, Carr GJ (1999) Understanding single-species and model ecosystem sensitivity: data-based comparison. Environ Toxicol Chem 18:1329–1346

Victor R, Ogbeibu AE (1986) Recolonization of macrobenthic invertebrates in a Nigerian stream after pesticide treatment and associated disruption. Environ Pollut Ser A Ecol Biol 41:125–137

Wagner C, Løkke H (1991) Estimation of ecotoxicological protection levels from NOEC toxicity data. Water Res 25:1237–1242

Wallace JB, Vogel DS, Cuffney TF (1986) Recovery of a headwater stream from an insecticide-induced community disturbance. J N Am Benthol Soc 5:115–126

Warner K, Fenderson OC (1962) Effects of DDT spraying for forest insects on Maine trout streams. J Wildl Manage 26:86–93

Wheelock CE, Eder KJ, Werner I, Huang HZ, Jones PD, Brammell BF, Elskus AA, Hammock BD (2005) Individual variability in esterase activity and CYP1A levels in Chinook salmon (*Oncorhynchus tshawytscha*) exposed to esfenvalerate and chlorpyrifos. Aquat Toxicol 74: 172–192

Whiles MR, Wallace JB (1995) Macroinvertebrate production in a headwater stream during recovery from anthropogenic disturbance and hydrologic extremes. Can J Fish Aquat Sci 52:2402–2422

Whitehouse P, Crane M, Grist E, O'Hagan A, Sorokin N (2004) Derivation and expression of water quality standards; opportunities and constraints in adopting risk-based approaches in EQS setting, UK

Wu JG, Laird DA (2004) Interactions of chlorpyrifos with colloidal materials in aqueous systems. J Environ Qual 33:1765–1770

Yount JD, Niemi GJ (1990) Recovery of lotic communities and ecosystems from disturbance – a narrative review of case studies. Environ Manage 14:547–569

Zabel TF, Cole S (1999) The derivation of environmental quality standards for the protection of aquatic life in the UK. J Chart Inst Water Environ Manag 13:436–440

Zischke JA, Arthur JW, Hermanutz RO, Hedtke SF, Helgen JC (1985) Effects of pentachlorophenol on invertebrates and fish in outdoor experimental channels. Aquat Toxicol 7:37–58

Index

A
Abbreviations, used in this volume (table), 4
Acetylcholinesterase (AChE) inhibition, chlorpyrifos, 11
Acronyms, used in this volume (table), 4
ACRs, *see* Acute-to-chronic ratios (ACRs)
Acute aquatic criteria derivation, SSD procedure, 106
Acute-to-chronic ratios (ACRs)
   calculation, pesticide default values (table), 69
   chronic criteria derivation, 114
   in criteria derivation, 66
   literature default values, 116
   single-chemical multispecies, 67
Acute criterion derivation
   AF procedure, 113
   UCDM, 105
Acute data, to estimate chronic toxicity, 16
Acute factor utilization, AF procedures, 64
Acute toxicity, pesticide data sets (table), 39
Additivity models, toxicity of mixtures, 81
AF, *see* Assessment factor (AF)
Air criteria
   criteria setting, 92
   harmonization, 92
      with aquatic criteria, 132
Aldrin
   data-set distribution test (diag.), 41
   toxicity data, comparative distribution fit (illus.), 45
Allowable exceedance, conclusion, 75
Alternative methods, for calculating criteria, 37
Antagonism, toxicity of mixtures, 84
Aquatic criteria
   development, needed ecotoxicity data, 32
   estimation technique, 15
   exposure-time parameter, 14
   harmonization
      with air values, 132
      with sediment values, 132
   role, multi-pathway exposure, 14
   setting, toxicity data summary (table), 21
   species data requirements, 8
Aquatic field data, evaluation and use, 102
Aquatic lab data, rating method (tables), 27, 28
Aquatic life criteria
   data requirements (table), 7
   ecosystem protection, 6
   toxicity data required, 7
Aquatic life water quality, pesticides, 1$f\!f$
Aquatic outdoor data, evaluation and use, 102
Aquatic outdoor field data, quality rating scheme (tables), 30, 31
Aquatic species
   criteria derivation data needs, 33
   ecotoxicity testing, 9
   exposure data evaluation, 102
Aquatic toxicity data, UCDM, 98
Assessment factor (AF)
   choosing toxicity values, 63
   procedures
      acute criterion derivation, 113
      acute factor utilization, 64
      criteria derivation, 60
   proper use, 62
   setting magnitude factors, 64
   used in existing methodologies (table), 61
Assumptions
   to derived criteria, review, 133
   made, UCDM, 93
Atrazine
   data-set distribution test (diag.), 42
   toxicity data, comparative distribution fit (illus.), 49
Averaging periods, criteria derivation, 69

B

Bioaccumulation
   role in criteria setting, 127
   UCDM and food residues, 89
Bioavailability
   criteria-setting implications, 76
   effect, criteria compliance, 118
Biomagnification factor (BMF), default values (table), 91
Burr III distribution fit, pesticide toxicity data (table), 50
Burr Type III distribution fit, pesticide data sets (table), 51

C

Chemical mixtures
   criteria-setting implications, 80
   effects, criteria compliance, 119
Chemicals listed in this volume, chemical names (table), 128
Chlordane, data-set distribution test (diag.), 42
Chlordane toxicity data, comparative distribution fit (illus.), 47
Chlorpyrifos
   AChE inhibition, 11
   aquatic criteria, calculated with UCDM (table), 137
   data-set distribution test (diag.), 41
   toxicity data, comparative distribution fit (illus.), 43
   water quality parameters, 3
Chronic criteria derivation
   herbicides, 117
   SSD procedure, 114
   using ACRs, 114
Chronic-data gaps, estimation technique, 103
Chronic toxicity estimates, from acute data, 16
Concentration addition model, toxicity of mixtures, 81
Criteria calculation
   approaches, 37
   SSD method, 38
Criteria compliance
   bioavailability effect, 118
   incorporating water quality effects, 117
   mixture effects, 119
   temperature and pH effects, 121
Criteria derivation
   AF procedure, 60
   averaging periods, 69
   data needs, aquatic species, 33
   data
      outliers, 54
      reduction methods, 35
      requirements (table), 7
   field data role, 13
   flow chart, UCDM (diag.), 97
   improvement, data generation, 94
   methods, evaluating ecotoxicity data, 25
   national water quality, 2
   rating single-species data quality (table), 26
   role of ACRs, 66
   SSD percentile cutoff point, 52
   taxa aggregation, 54
   using default ACRs, 116
   using multispecies, 13
   water quality effects, 76
Criteria setting
   chemical mixtures, 80
   ecotoxicity data checking, 88
   implications, bioavailability, 76
   5th percentile calculation, 112
   role
      bioaccumulation, 127
      of ecosystem data, 125
      of TES, 125
      secondary toxicity, 127
   sediment harmonization, 92
   sensitive species protection, 124
   SSD flow chart (diag.), 111
Criteria validation, against ecotoxicity data, 124

D

Data
   collection details, UCDM, 98
   flow chart, UCDM (diag.), 97
   -gap filling, estimation techniques, 15
   generation, criteria derivation improvement, 94
   outliers, in criteria derivation, 54
   quality, ecotoxicity data summaries, 20
   quantity required, ecotoxicity, 32
   reduction methods, criteria derivation, 35
   relevance and reliability scores, data categories (table), 31
   requirements
      criteria derivation (table), 7
   UCDM, 6
   sources for UCDM development (table), 17
DDT
   data-set distribution test (diag.), 41
   toxicity data, comparative distribution fit (illus.), 43
Default ACR calculation, pesticides (table), 69
Definitions, UCDM, 95

Diazinon
    data-set distribution test (diag.), 42
    toxicity data, comparative distribution fit
        (illus.), 48
Dieldrin
    data-set distribution test (diag.), 42
    toxicity data, comparative distribution fit
        (illus.), 46

E
Ecosystem
    data, role in criteria setting, 125
    protection, aquatic life criteria, 6
Ecotoxicity
    evaluation, hypothesis tests vs. regression
        analysis, 9
    hypothesis testing issues, 9
    regression analysis, 10
Ecotoxicity data
    checking, criteria setting, 88
    criteria validation, 124
    evaluation
        criteria derivation methods, 25
        UCDM, 101
        and use, single- and multi-species, 101
    quantity required, 32
    required, UCDM, 99
    summaries, quality ratings, 20
Endocrine disruption, endpoints assessed, 11
Endosulfan
    data-set distribution test (diag.), 42
    toxicity data, comparative distribution fit
        (illus.), 47
Endpoints, derivation methods context, 11
Endrin
    data-set distribution test (diag.), 41
    toxicity data, comparative distribution fit
        (illus.), 44
Estimation techniques, filling chronic data
    gaps, 103
Exceedance limits, frequency, 71

F
Field data, in criteria derivation, 13
Frequency of exceedance, limits, 71

G
Goals, UCDM, 95

H
Harmonization for sediments, criteria
    setting, 92
Heptachlor
    data-set distribution test (diag.), 42
    toxicity data, comparative distribution fit
        (illus.), 46
Herbicides, deriving chronic criteria, 117
Human health data, role in UCDM, 101
Hypothesis testing
    ecotoxicity evaluation, 9
    issues, ecotoxicity, 9

I
Interspecies correlations, QSAR use, 16

L
Limitations
    to derived criteria, review, 133
    UCDM, 93
Lindane
    data-set distribution test (diag.), 41
    toxicity data, comparative distribution fit
        (illus.), 45
Log-normal distribution fit, pesticide data sets
    (table), 51
Log-triangular distribution fit, pesticide data
    sets (table), 51

M
Magnitude factors, AF context, 64
Mesocosm data, evaluation and use, 102
Method comparisons, analyzing pesticide data
    sets, 55
Methodologies, current AFs used (table), 61
Method for rating quality
    aquatic outdoor field data (tables), 30, 31
    model ecosystem data (table), 30
    single-species aquatic data (tables), 27, 28
    terrestrial lab data (table), 31
Methods
    comparison, pesticide data analysis
        (table), 55
    for determining
        octanol-water partitioning coefficients
            (table), 25
        physical-chemical parameters
            (table), 24
    water quality criteria, 1ff
Microcosm data, evaluation and use, 102
Model ecosystem data
    evaluation and use, 102
    quality rating scheme (table), 30
Models, SSD methods, 58
Multi-pathway exposures, aquatic criteria
    role, 14
Multispecies
    data, role in UCDM, 100
    use, in criteria derivation, 13

O

Octanol-water partition coefficient, acceptable development methods (table), 25

P

Percentile factors, pesticide data sets (table), 65
Pesticide data analysis, methods comparison (table), 55
Pesticide data sets
  acute toxicity (table), 39
  analysis, method comparisons, 55
  Burr Type III distribution fit (table), 51
  log-normal distribution fit (table), 51
  log-triangular distribution fit (table), 51
  percentile factors (table), 65
Pesticides
  Burr III family distribution fit (table), 50
  listed in this volume, chemical names (table), 128
  water quality criteria, 1ff
pH effects, criteria compliance, 121
Physical-chemical data
  differential quality, 22
  evaluation, UCDM, 101
  generation, acceptable methods (table), 24
  water quality criteria use, 22
Population level
  effects, key endpoints, 12
  endpoints, use in UCDM, 13

Q

Quantitative structure activity relationships (QSAR)
  approach, data-gap filling, 15
  interspecies correlations, 16
  toxicity estimation procedure (table), 126

R

Ranking scheme, data relevance and reliability (table), 31
Regression analysis, ecotoxicity evaluation, 9, 10
Reproductive endpoints, ecotoxicity assessment, 11
Response addition model, toxicity of mixtures, 83

S

Secondary poisoning
  role in criteria setting, 127
  wildlife dietary intake, 90
Sediment criteria, harmonization with aquatic values, 132
  harmonization, in criteria setting, 92
Sensitive species protection, setting criteria, 124
Single species
  aquatic data, rating method (table), 27
  data, for criteria derivation (table), 26
Species data requirements, aquatic criteria, 8
Species sensitivity distribution (SSD)
  for calculating criteria, 38
  failure, fit test, 111
  flow chart, in criteria setting (diag.), 111
  model comparison, discussion, 58
  percentile cutoff point, criteria derivation, 52
  procedure
    checking goodness of fit, 109
    chronic criterion derivation, 114
    setting confidence levels, 53
    taxa aggregation, 54
  UCDM context, 59
  use, deriving acute criteria, 106
Synergism, toxicity of mixtures, 84

T

Taxa aggregation
  criteria derivation, 54
  SSD procedures, 54
Temperature effects, criteria compliance, 121
Terrestrial data, role in UCDM, 100
Terrestrial lab data, quality rating scheme, 31
Threatened and endangered species (TES)
  criteria setting, 88
  role in criteria setting, 125
Toxaphene
  data-set distribution test (diag.), 41
  toxicity data, comparative distribution fit (illus.), 44
Toxic endpoints assessed, endocrine disruption, 11
Toxicity
  estimation procedure, QSARs (table), 126
  implications, chemical mixtures, 80
  UCDM requirements, 10
Toxicity data
  aquatic species required, 8
  fit to Burr III, pesticides (table), 50
  required, aquatic life criteria, 7
  summary, aquatic-criteria setting (table), 21
  time parameters, 14
Toxicity endpoints
  aquatic species, 8

Index

assessed, aquatic life criteria, 11
population-level effects, 12
Toxicity of mixtures
    additivity models, 81
    combining models, 84
    concentration addition model, 81
    response addition model, 83
    synergism and antagonism, 84
Toxicity values
    normalization procedure, water quality, 122
    use with AFs, 63

U

UCDM, *see* University of California-Davis methodology (UCDM)
Uncertainties, UCDM, 93
University of California-Davis methodology (UCDM)
    acute criterion derivation, 105
    aquatic toxicity data, 98
    assumptions and limitations, 93
    bioaccumulation and food residues, 89
    chlorpyrifos aquatic criteria values (table), 137
    context, SSDs, 59
    criteria derivation flow chart (diag.), 97
    data
        collections details, 98
        flow chart (diag.), 97
        requirements, 6
    development

data sources (table), 17
web-address sources (table), 19
evaluating ecotoxicity data, 101
evaluating physical-chemical data, 101
goals
    and definitions, 95
    water quality criteria, 5
population-level endpoints, 13
requirements, acute and chronic toxicity, 10
uncertainties, 93
water quality, 3
University of California, water-quality methodology, 1*ff*
USEPA, water quality methods, 2

W

Water quality
    methods, USEPA, 2
    toxicity value normalization procedure, 122
Water quality criteria
    methodology goals, 5
    pesticides, 1*ff*
    UCDM, 3
    use, physical-chemical data, 22
Water quality effects
    criteria compliance, 117
    role in UCDM, 100
    on toxicity, 76
Web-address sources, UCDM development (table), 19
Wildlife dietary intake, secondary poisoning, 90

Breinigsville, PA USA
01 October 2010
246476BV00005B/9/P